Metric Learning

Synthesis Lectures on Artificial Intelligence and Machine Learning

Metric Learning
Aurélien Bellet, Amaury Habrard, and Marc Sebban
2015

Graph-Based Semi-Supervised Learning
Amarnag Subramanya and Partha Pratim Talukdar
2014

Robot Learning from Human Teachers
Sonia Chernova and Andrea L. Thomaz
2014

General Game Playing
Michael Genesereth and Michael Thielscher
2014

Judgment Aggregation: A Primer
Davide Grossi and Gabriella Pigozzi
2014

An Introduction to Constraint-Based Temporal Reasoning
Roman Barták, Robert A. Morris, and K. Brent Venable
2014

Reasoning with Probabilistic and Deterministic Graphical Models: Exact Algorithms
Rina Dechter
2013

Introduction to Intelligent Systems in Traffic and Transportation
Ana L.C. Bazzan and Franziska Klügl
2013

A Concise Introduction to Models and Methods for Automated Planning
Hector Geffner and Blai Bonet
2013

Essential Principles for Autonomous Robotics
Henry Hexmoor
2013

Case-Based Reasoning: A Concise Introduction
Beatriz López
2013

Answer Set Solving in Practice
Martin Gebser, Roland Kaminski, Benjamin Kaufmann, and Torsten Schaub
2012

Planning with Markov Decision Processes: An AI Perspective
Mausam and Andrey Kolobov
2012

Active Learning
Burr Settles
2012

Computational Aspects of Cooperative Game Theory
Georgios Chalkiadakis, Edith Elkind, and Michael Wooldridge
2011

Representations and Techniques for 3D Object Recognition and Scene Interpretation
Derek Hoiem and Silvio Savarese
2011

A Short Introduction to Preferences: Between Artificial Intelligence and Social Choice
Francesca Rossi, Kristen Brent Venable, and Toby Walsh
2011

Human Computation
Edith Law and Luis von Ahn
2011

Trading Agents
Michael P. Wellman
2011

Visual Object Recognition
Kristen Grauman and Bastian Leibe
2011

Learning with Support Vector Machines
Colin Campbell and Yiming Ying
2011

Algorithms for Reinforcement Learning
Csaba Szepesvári
2010

Data Integration: The Relational Logic Approach
Michael Genesereth
2010

Markov Logic: An Interface Layer for Artificial Intelligence
Pedro Domingos and Daniel Lowd
2009

Introduction to Semi-Supervised Learning
XiaojinZhu and Andrew B.Goldberg
2009

Action Programming Languages
Michael Thielscher
2008

Representation Discovery using Harmonic Analysis
Sridhar Mahadevan
2008

Essentials of Game Theory: A Concise Multidisciplinary Introduction
Kevin Leyton-Brown and Yoav Shoham
2008

A Concise Introduction to Multiagent Systems and Distributed Artificial Intelligence
Nikos Vlassis
2007

Intelligent Autonomous Robotics: A Robot Soccer Case Study
Peter Stone
2007

Metric Learning

Aurélien Bellet, Amaury Habrard, and Marc Sebban

ISBN: 978-3-031-00444-5 paperback
ISBN: 978-3-031-01572-4 ebook

DOI 10.1007/978-3-031-01572-4

A Publication in the Springer series
SYNTHESIS LECTURES ON ARTIFICIAL INTELLIGENCE AND MACHINE LEARNING

Lecture #30
Series Editors: Ronald J. Brachman, *Yahoo! Labs*
William W. Cohen, *Carnegie Mellon University*
Peter Stone, *University of Texas at Austin*
Series ISSN
Print 1939-4608 Electronic 1939-4616

Metric Learning

Aurélien Bellet
Télécom ParisTech

Amaury Habrard
Université de Saint-Etienne

Marc Sebban
Université de Saint-Etienne

SYNTHESIS LECTURES ON ARTIFICIAL INTELLIGENCE AND MACHINE LEARNING #30

ABSTRACT

Similarity between objects plays an important role in both human cognitive processes and artificial systems for recognition and categorization. How to appropriately measure such similarities for a given task is crucial to the performance of many machine learning, pattern recognition and data mining methods. This book is devoted to metric learning, a set of techniques to automatically learn similarity and distance functions from data that has attracted a lot of interest in machine learning and related fields in the past ten years. In this book, we provide a thorough review of the metric learning literature that covers algorithms, theory and applications for both numerical and structured data. We first introduce relevant definitions and classic metric functions, as well as examples of their use in machine learning and data mining. We then review a wide range of metric learning algorithms, starting with the simple setting of linear distance and similarity learning. We show how one may scale-up these methods to very large amounts of training data. To go beyond the linear case, we discuss methods that learn nonlinear metrics or multiple linear metrics throughout the feature space, and review methods for more complex settings such as multi-task and semi-supervised learning. Although most of the existing work has focused on numerical data, we cover the literature on metric learning for structured data like strings, trees, graphs and time series. In the more technical part of the book, we present some recent statistical frameworks for analyzing the generalization performance in metric learning and derive results for some of the algorithms presented earlier. Finally, we illustrate the relevance of metric learning in real-world problems through a series of successful applications to computer vision, bioinformatics and information retrieval.

KEYWORDS

metric learning, similarity learning, Mahalanobis distance, edit distance, structured data, learning theory

Contents

CHAPTER 1

Introduction

Many cognitive processes, such as recognition and categorization, are assumed to require forming similarity and distance judgments between perceptual or conceptual representations. Essentially, when facing stimuli or situations similar to what we have encountered before, we expect similar responses and take similar actions. This has led psychologists to develop a variety of cognitive theories and mathematical models of similarity [Ashby and Perrin, 1988, Hahn et al., 2003, Markman and Gentner, 1993, Medin et al., 1993, Nosofsky, 1986, Shepard, 1987, Tversky, 1977]. Given its intuitive appeal as an explanatory notion, it comes as no surprise that the concept of similarity underlies most of the machine learning, pattern recognition and data mining techniques. Prominent examples include nearest-neighbor classification [Cover and Hart, 1967], data clustering [Lloyd, 1982], kernel methods [Schölkopf and Smola, 2001] and many information retrieval methods [Manning et al., 2008].

A question of crucial importance to the performance of the above methods is how to appropriately measure the similarity or distance between objects. Assuming that they are represented by a set of numerical features, the Euclidean distance can be used but often leads to suboptimal performance, arguably because of its inability to take into account the relevance degree of different features. This hypothesis is supported by psychological studies that suggest humans weight features differently depending on the context [Goldstone et al., 1997, Nosofsky, 1986]. To illustrate this, consider a database of face images. If the goal is to do facial identification, the similarity between two images should be based on features such as hair color, face shape and proportions, etc. On the other hand, if the goal is to determine the facial expression, the similarity should rather focus on a different set of features, such as position of eyebrows, mouth, cheeks, etc. In practice, it is difficult to define the optimal similarity measure for a particular task, even for a domain expert.

It is often much easier to collect pairs of objects that are similar/dissimilar, or triplets of the form "x is more similar to y than to z." Metric learning consists in automatically learning how to appropriately measure similarity and distance from this type of side information. Although its origins can be traced back to some earlier work [e.g., Baxter and Bartlett, 1997, Friedman, 1994, Fukunaga, 1990, Hastie and Tibshirani, 1996, Short and Fukunaga, 1981], this line of research truly emerged with the pioneering work of Xing et al. [2002] that formulates metric learning as a convex optimization problem. It has since been a hot research topic, with numerous papers published at top machine learning, computer vision and data mining venues, as well as the organization of several tutorials (ICML 2010, ECCV 2010, SDM 2012) and workshops (ICCV 2011, NIPS 2011, ICML 2013).

1.1 METRIC LEARNING IN A NUTSHELL

The goal of metric learning is to adapt some pairwise real-valued *metric*,[1] say the Mahalanobis distance $d_M(\boldsymbol{x}, \boldsymbol{x}') = \sqrt{(\boldsymbol{x} - \boldsymbol{x}')^T \boldsymbol{M}(\boldsymbol{x} - \boldsymbol{x}')}$, to the problem of interest using the information brought by training examples. Most methods learn the metric from side information of the following form:

- Must-link / cannot-link constraints (sometimes called positive / negative pairs):

$$\begin{aligned}\mathcal{S} &= \{(x_i, x_j) : x_i \text{ and } x_j \text{ should be similar}\},\\ \mathcal{D} &= \{(x_i, x_j) : x_i \text{ and } x_j \text{ should be dissimilar}\}.\end{aligned}$$

- Relative constraints (sometimes called training triplets):

$$\mathcal{R} = \{(x_i, x_j, x_k) : x_i \text{ should be more similar to } x_j \text{ than to } x_k\}.$$

A metric learning algorithm basically aims at finding the parameters of the metric (here, the matrix $\boldsymbol{M}$) such that it best agrees with these constraints, in an effort to approximate the underlying semantic metric (see Figure 1.1 for an illustration).

1.2 RELATED TOPICS

We briefly present three research topics that are related to metric learning but are outside the scope of this book.

Kernel learning While metric learning is parametric (one learns the parameters of a given form of metric), kernel learning is usually nonparametric: the kernel matrix is learned without any assumption on the form of the kernel that implicitly generated it. These approaches are thus powerful but limited to the transductive setting: the resulting kernel is difficult to use on new data. The interested reader may refer to the recent survey on kernel learning by Abbasnejad et al. [2012].

Multiple kernel learning Unlike kernel learning, Multiple Kernel Learning (MKL) is parametric: it learns a combination of predefined base kernels. In this regard, it can be seen as more restrictive than metric learning, but as opposed to kernel learning, MKL can be applied in the inductive setting. The interested reader may refer to the recent survey on MKL by Gönen and Alpaydin [2011].

[1]By a common abuse of terminology, we will use the term *metric* to refer to any pairwise function measuring a distance or similarity between objects.

[2]`http://www.vision.caltech.edu/html-files/archive.html`.

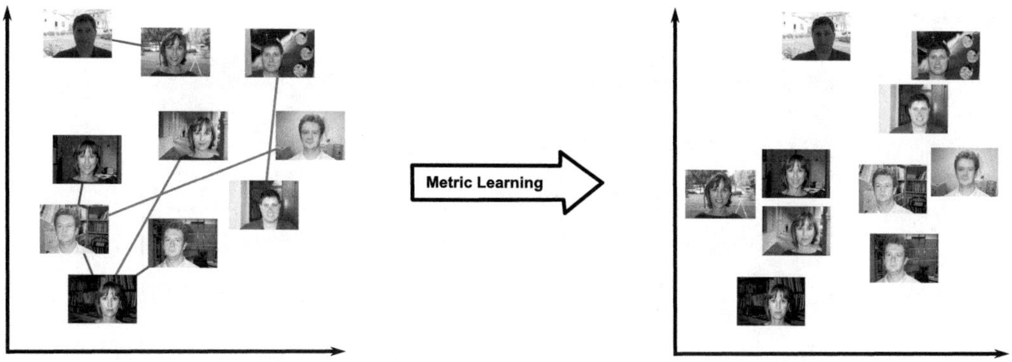

Figure 1.1: Illustration of metric learning applied to a face recognition task. For simplicity, images are represented as points in two dimensions. Pairwise constraints, shown in the left pane, are composed of images representing the same person (must-link, shown in green) or different persons (cannot-link, shown in red). We wish to adapt the metric so that there are fewer constraint violations (right pane). Images are taken from the Caltech Faces dataset.[2]

Dimensionality reduction Supervised dimensionality reduction aims at finding a low-dimensional representation that maximizes the separation of labeled data. As we shall see later, this has connections with metric learning, although the primary objective is different. Unsupervised dimensionality reduction, or manifold learning, usually assumes that the (unlabeled) data lie on an embedded low-dimensional manifold within the higher-dimensional space and aim at "unfolding" it. These methods aim at capturing or preserving some properties of the original data (such as the variance or local distance measurements) in the low-dimensional representation.[3] The interested reader may refer to the surveys by Fodor [2002] and van der Maaten et al. [2009].

1.3 PREREQUISITES AND NOTATIONS

We tried to make this book as self-contained as possible. Basic knowledge of convex optimization is required: the reader may refer to the classic convex optimization book of Boyd and Vandenberghe [2004] as well as the recent book on optimization for machine learning edited by Sra et al. [2011]. We also assume that the reader has some knowledge of linear algebra as well as some familiarity with probability theory, statistics and machine learning. The notations used throughout this book are summarized in Table 1.1.

[3]These approaches are sometimes referred to as "unsupervised metric learning," which is somewhat misleading because they do not optimize a notion of metric.

Table 1.1: Summary of the main notations.

Notation	Description
$\mathbb{R}$	Set of real numbers
$\mathbb{R}^d$	Set of d-dimensional real-valued vectors
$\mathbb{R}^{c\times d}$	Set of $c \times d$ real-valued matrices
$\mathbb{S}_+^d$	Cone of symmetric PSD $d \times d$ real-valued matrices
$\mathcal{X}$	Input (instance) space
$\mathcal{Y}$	Output (label) space
$\mathcal{S}$	Set of must-link constraints
$\mathcal{D}$	Set of cannot-link constraints
$\mathcal{R}$	Set of relative constraints
$z = (x, y) \in \mathcal{X} \times \mathcal{Y}$	An arbitrary labeled instance
$\boldsymbol{x}$	An arbitrary vector
$\boldsymbol{M}$	An arbitrary matrix
$\boldsymbol{I}$	Identity matrix
$\boldsymbol{M} \succeq 0$	PSD matrix $\boldsymbol{M}$
$\lVert \cdot \rVert_p$	p-norm
$\lVert \cdot \rVert_{\mathcal{F}}$	Frobenius norm
$\lVert \cdot \rVert_*$	Nuclear norm
$\text{tr}(\boldsymbol{M})$	Trace of matrix $\boldsymbol{M}$
$[t]_+ = \max(0, t)$	Hinge loss function
ξ	Slack variable
Σ	Finite alphabet
x	String of finite size

1.4 OUTLINE

The book proposes a wide overview of the metric learning literature, covering algorithms, theory and applications for both numerical and structured data.

We begin with the introduction of preliminary notions. Chapter 2 is devoted to metrics, with important definitions, examples of popular metrics for numerical and structured data, and a brief summary of their use in machine learning and data mining. Chapter 3 introduces five key properties of metric learning algorithms that form the basis of a taxonomy of the algorithms studied in this book (summarized in Chapter 10).

The next four chapters of the book take the form of a review of existing metric learning algorithms. We take a tutorial-like approach: we focus on the most important methods and promising recent trends, describing them in much detail. In that sense, the book is complementary to our recent survey [Bellet et al., 2013] which covers a wider range of work in less depth. Chapter 4 is devoted to learning linear metrics, some of them in the form of proper distance functions, others in the form of arbitrary similarity functions that do not satisfy the distance axioms. We also discuss how to scale-up metric learning to the large-scale setting, where the number of training samples and/or the dimensionality of the data is very large. Chapter 5 investigates how one can go beyond learning simple linear metrics to improve the performance. In particular, we present strategies to make a linear metric learning algorithm nonlinear. We also cover methods that directly learn nonlinear metrics, as well as those learning multiple local metrics that vary across the feature space. Chapter 6 presents extensions of metric learning to a few special settings: multi-task and transfer learning, ranking, semi-supervised learning, domain adaptation and histogram data. Chapter 7 deals with metric learning for structured data such as strings, trees, graphs and time series. We mainly focus on methods that learn metrics in the form of edit distances or structural/temporal alignments.

Chapter 8 is devoted to the question of generalization in metric learning. Relying on statistical learning theory, the goal is to derive formal guarantees for the performance of the learned metric on unseen data. We introduce several useful frameworks (e.g., uniform stability, algorithmic robustness) and show how they can be used to obtain generalization results for a wide range of metric learning algorithms. We show that one may not only obtain guarantees for the consistency of learned metric itself, but also in some cases for its performance in a classification task.

Chapter 9 illustrates the fact that metric learning has been successfully applied to many real-world problems. We review some important applications in the fields of computer vision, bioinformatics and information retrieval.

Finally, Chapter 10 provides a brief summary along with a taxonomy of the algorithms studied in the book, and attempts to draw promising lines for future research.

CHAPTER 2

Metrics

This chapter introduces some background knowledge on metrics and their applications. Section 2.1 provides definitions for distance, similarity and kernel functions. Some standard metrics are presented in Section 2.2. We conclude this chapter by briefly discussing the use of metrics in machine learning and data mining in Section 2.3.

2.1 GENERAL DEFINITIONS

A distance function satisfies four essential conditions, that we recall below.

Definition 2.1 A *distance* over a set $\mathcal{X}$ is a pairwise function $d : \mathcal{X} \times \mathcal{X} \to \mathbb{R}$ which satisfies the following conditions $\forall x, x', x'' \in \mathcal{X}$:

1. $d(x, x') \geq 0$ (nonnegativity),
2. $d(x, x') = 0$ if and only if $x = x'$ (identity of indiscernibles),
3. $d(x, x') = d(x', x)$ (symmetry),
4. $d(x, x'') \leq d(x, x') + d(x', x'')$ (triangle inequality).

These conditions, referred to as the *distance axioms*, reflect intuitive properties about the concept of distance. A *pseudo-distance* is required to satisfy these properties, except that instead of condition 2, only $d(x, x) = 0$ needs to be satisfied.

While a distance function is a well-defined mathematical object, there is no general agreement on the definition of a similarity function, which can essentially be any pairwise function. We will use the following definition.

Definition 2.2 A *similarity function* is a pairwise function $S : \mathcal{X} \times \mathcal{X} \to \mathbb{R}$. We say that S is a symmetric similarity function if $\forall x, x' \in \mathcal{X}$, $S(x, x') = S(x', x)$.
By convention, the larger the value of a similarity function, the more similar the pair of inputs (otherwise we will use the term *dissimilarity function*). Trivially, distance functions are dissimilarity functions.

We also introduce the notion of kernel.

Definition 2.3 A symmetric similarity function K is a *kernel* if there exists a mapping function $\phi : \mathcal{X} \to \mathbb{H}$ from the instance space $\mathcal{X}$ to a vector space $\mathbb{H}$ equipped with an inner product $\langle \cdot, \cdot \rangle$

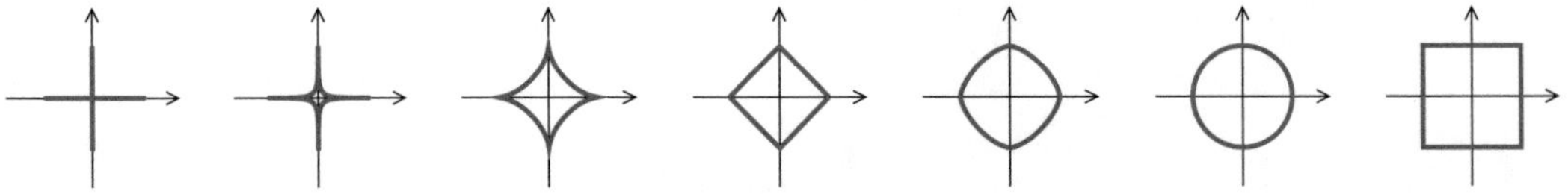

Figure 2.1: Minkowski distances: unit circles for various values of p.

($\mathbb{H}$ is called a Hilbert space) such that K can be written as

$$K(x, x') = \langle \phi(x), \phi(x') \rangle .$$

Equivalently, K is a kernel if it is positive semi-definite (PSD), i.e.,

$$\sum_{i=1}^{n} \sum_{j=1}^{n} c_i c_j K(x_i, x_j) \geq 0$$

for all finite sequences of $x_1, \ldots, x_n \in \mathcal{X}$ and $c_1, \ldots, c_n \in \mathbb{R}$.
In other words, a kernel is a function that takes the form of an inner product in $\mathbb{H}$, and can thus be interpreted as a measure of similarity [Schölkopf and Smola, 2001].

2.2 COMMONLY USED METRICS

In this section, we give some examples of popular metrics.

2.2.1 METRICS FOR NUMERICAL DATA

We first focus on the case where data points lie in a vector space $\mathcal{X} \subseteq \mathbb{R}^d$.

Minkowski distances Minkowski distances are a family of distances induced by L_p norms. Formally, for $p \geq 1$,

$$d_p(\boldsymbol{x}, \boldsymbol{x}') = \|\boldsymbol{x} - \boldsymbol{x}'\|_p = \left(\sum_{i=1}^{d} |x_i - x_i'|^p \right)^{1/p} . \tag{2.1}$$

Figure 2.1 shows the corresponding unit circles for several values of p. From (2.1) we can recover three widely used distances:

- When $p = 1$, we get the Manhattan distance:

$$d_{man}(\boldsymbol{x}, \boldsymbol{x}') = \|\boldsymbol{x} - \boldsymbol{x}'\|_1 = \sum_{i=1}^{d} |x_i - x_i'|.$$

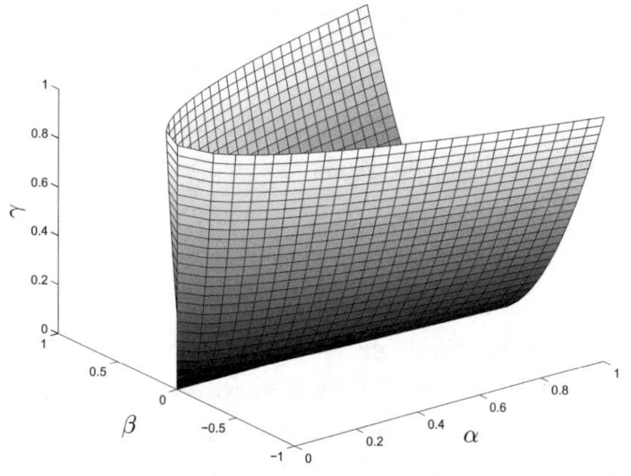

Figure 2.2: The cone $\mathbb{S}_+^2$ of symmetric positive semi-definite 2x2 matrices of the form $\begin{bmatrix} \alpha & \beta \\ \beta & \gamma \end{bmatrix}$.

- When $p = 2$, (2.1) becomes the "ordinary" Euclidean distance:

$$d_{euc}(\boldsymbol{x}, \boldsymbol{x}') = \|\boldsymbol{x} - \boldsymbol{x}'\|_2 = \left(\sum_{i=1}^{d} |x_i - x_i'|^2 \right)^{1/2} = \sqrt{(\boldsymbol{x} - \boldsymbol{x}')^T (\boldsymbol{x} - \boldsymbol{x}')}.$$

- When $p \to \infty$, we get the Chebyshev distance:

$$d_{che}(\boldsymbol{x}, \boldsymbol{x}') = \|\boldsymbol{x} - \boldsymbol{x}'\|_\infty = \max_i |x_i - x_i'|.$$

Note that when $0 < p < 1$, d_p is not a proper distance (it violates the triangle inequality).

Mahalanobis distances The term Mahalanobis distance comes from Mahalanobis [1936] and originally refers to a distance measure that incorporates the correlation between features:

$$d_{\boldsymbol{\Sigma}^{-1}}(\boldsymbol{x}, \boldsymbol{x}') = \sqrt{(\boldsymbol{x} - \boldsymbol{x}')^T \boldsymbol{\Sigma}^{-1} (\boldsymbol{x} - \boldsymbol{x}')},$$

where $\boldsymbol{x}$ and $\boldsymbol{x}'$ are random vectors from the same distribution with covariance matrix $\boldsymbol{\Sigma}$.

By an abuse of terminology common in the metric learning literature, we will use Mahalanobis distance to refer to generalized quadratic distances, defined as

$$d_{\boldsymbol{M}}(\boldsymbol{x}, \boldsymbol{x}') = \sqrt{(\boldsymbol{x} - \boldsymbol{x}')^T \boldsymbol{M} (\boldsymbol{x} - \boldsymbol{x}')}$$

and parameterized by $\boldsymbol{M} \in \mathbb{S}_+^d$, where $\mathbb{S}_+^d$ is the cone of symmetric positive semi-definite (PSD) $d \times d$ real-valued matrices (see Figure 2.2). The condition $\boldsymbol{M} \in \mathbb{S}_+^d$ ensures that $d_{\boldsymbol{M}}$ is a pseudo-distance.

When $\boldsymbol{M}$ is the identity matrix, we recover the Euclidean distance. Otherwise, one can express $\boldsymbol{M}$ as $\boldsymbol{L}^T\boldsymbol{L}$, where $\boldsymbol{L} \in \mathbb{R}^{k\times d}$ where k is the rank of $\boldsymbol{M}$. We can then rewrite $d_{\boldsymbol{M}}(\boldsymbol{x}, \boldsymbol{x}')$ as follows:

$$\begin{aligned} d_{\boldsymbol{M}}(\boldsymbol{x}, \boldsymbol{x}') &= \sqrt{(\boldsymbol{x}-\boldsymbol{x}')^T\boldsymbol{M}(\boldsymbol{x}-\boldsymbol{x}')} \\ &= \sqrt{(\boldsymbol{x}-\boldsymbol{x}')^T\boldsymbol{L}^T\boldsymbol{L}(\boldsymbol{x}-\boldsymbol{x}')} \\ &= \sqrt{(\boldsymbol{L}\boldsymbol{x}-\boldsymbol{L}\boldsymbol{x}')^T(\boldsymbol{L}\boldsymbol{x}-\boldsymbol{L}\boldsymbol{x}')}. \end{aligned}$$

Thus, a Mahalanobis distance implicitly corresponds to computing the Euclidean distance after the linear projection of the data defined by the transformation matrix $\boldsymbol{L}$. Note that if $\boldsymbol{M}$ is low-rank, i.e., rank$(\boldsymbol{M}) = k < d$, then it induces a linear projection of the data into a space of lower dimension r. It thus allows a more compact representation of the data and cheaper distance computations, especially when the original feature space is high-dimensional. These nice properties explain why Mahalanobis distances have attracted a lot of interest in metric learning, as we shall see in Chapter 4.

Cosine similarity The cosine similarity measures the cosine of the angle between its inputs, and can be computed as

$$S_{cos}(\boldsymbol{x}, \boldsymbol{x}') = \frac{\boldsymbol{x}^T\boldsymbol{x}'}{\|\boldsymbol{x}\|_2\|\boldsymbol{x}'\|_2}.$$

The cosine similarity is widely used in data mining, in particular in text retrieval [Baeza-Yates and Ribeiro-Neto, 1999] and more recently in image retrieval [see for instance Sivic and Zisserman, 2009] when data are represented as term vectors [Salton et al., 1975].

Bilinear similarity The bilinear similarity is related to the cosine similarity but does not include normalization by the norms of the inputs and is parameterized by a matrix $\boldsymbol{M}$:

$$S_{\boldsymbol{M}}(\boldsymbol{x}, \boldsymbol{x}') = \boldsymbol{x}^T\boldsymbol{M}\boldsymbol{x}',$$

where $\boldsymbol{M} \in \mathbb{R}^{d\times d}$ is typically not required to be PSD nor symmetric. The bilinear similarity has been used for instance in image retrieval [Deng et al., 2011]. When $\boldsymbol{M}$ is the identity matrix, $S_{\boldsymbol{M}}$ amounts to an unnormalized cosine similarity.

The bilinear similarity has two main advantages. First, it is efficiently computable for sparse inputs: if $\boldsymbol{x}$ and $\boldsymbol{x}'$ have respectively k and k' nonzero entries, $S_{\boldsymbol{M}}(\boldsymbol{x}, \boldsymbol{x}')$ can be computed in $O(k \cdot k')$ operations. Second, it can be used as a similarity measure between instances of different dimension (for example, a document and a query in web search) by using a nonsquare matrix $\boldsymbol{M}$.

Linear kernel The linear kernel is simply the inner product in the original space $\mathcal{X}$:

$$K_{lin}(\boldsymbol{x}, \boldsymbol{x}') = \langle\boldsymbol{x}, \boldsymbol{x}'\rangle = \boldsymbol{x}^T\boldsymbol{x}'.$$

In other words, the corresponding space $\mathbb{H} = \mathcal{X}$ and ϕ is an identity map: $\forall \boldsymbol{x} \in \mathcal{X}, \phi(\boldsymbol{x}) = \boldsymbol{x}$. Note that K_{lin} corresponds to the bilinear similarity with $\boldsymbol{M} = \boldsymbol{I}$.

Polynomial kernels The polynomial kernel of degree $p \in \mathbb{N}$ is defined as:

$$K_{poly}(\boldsymbol{x}, \boldsymbol{x}') = (\langle \boldsymbol{x}, \boldsymbol{x}' \rangle + 1)^p.$$

It can be shown that the space $\mathbb{H}$ corresponding to K_{poly} is the space of all monomials of degree up to p.

Gaussian RBF kernel The Gaussian RBF kernel is a widely used kernel function defined as

$$K_{rbf}(\boldsymbol{x}, \boldsymbol{x}') = \exp\left(-\frac{\|\boldsymbol{x} - \boldsymbol{x}'\|_2^2}{2\sigma^2}\right),$$

where $\sigma^2 > 0$ is called the width parameter. For this kernel, it can be shown that $\mathbb{H}$ is infinite-dimensional.

Histogram distances A (normalized) histogram is a feature vector on the probability simplex $\mathcal{S}^d = \{\boldsymbol{x} \in \mathbb{R}^d : \boldsymbol{x} \geq 0, \sum_i x_i = 1\}$. Such representation is common in text processing and computer vision, where documents are represented as a frequency vector of (visual) words [Li and Perona, 2005, Salton et al., 1975].

The χ^2 distance [Hafner et al., 1995] is a histogram pseudo-distance derived from the χ^2 statistical test and defined as[1]

$$\chi^2(\boldsymbol{x}, \boldsymbol{x}') = \frac{1}{2} \sum_{i=1}^{d} \frac{(x_i - x_i')^2}{x_i + x_i'}. \tag{2.2}$$

The χ^2 is bin-to-bin histogram distance. A popular cross-bin distance is the Earth Mover's Distance (EMD) introduced by Rubner et al. [2000]. Due to its rather complex form, we defer its exposition to Section 6.4.

2.2.2 METRICS FOR STRUCTURED DATA

We now turn to the case where data instances are structured objects, such as strings, trees or graphs. We first give a few definitions.

Definition 2.4 An *alphabet* Σ is a finite nonempty set of symbols.

Definition 2.5 A *string* x is a finite sequence of symbols from Σ. The empty string/symbol is denoted by \$ and Σ^* is the set of all finite strings (including \$) that can be generated from Σ. Finally, the length of a string x is denoted by $|\mathsf{x}|$.

Definition 2.6 Let T be a rooted tree. We call T a *labeled tree* if each node in T is assigned a symbol from an alphabet Σ. We call T an *ordered tree* if a left-to-right order among siblings in T is given.

[1]The sum in (2.2) must be restricted to entries that are nonzero in either $\boldsymbol{x}$ or $\boldsymbol{x}'$ to avoid division by zero.

Table 2.1: Example of edit cost matrix $\boldsymbol{C}$. Here, $\Sigma = \{\mathtt{a}, \mathtt{b}\}$.

C	\$	a	b
\$	0	2	10
a	2	0	4
b	10	4	0

Definition 2.7 A *graph* $G = (V, E)$ is composed of a set of vertices V and a set of edges $E \subseteq V \times V$, where $(e, e') \in E$ indicates that there is an edge between e and e'.

In the following, we introduce some commonly used metrics on structured data.

Hamming distance The Hamming distance is a distance between strings of identical length and is equal to the number of positions at which the symbols differ. It is used mostly for binary strings and is defined by

$$d_{ham}(\mathsf{x}, \mathsf{x}') = |\{i : \mathsf{x}_i \neq \mathsf{x}'_i\}|.$$

String edit distance The string edit distance [Levenshtein, 1966] is a distance between strings of possibly different length built from an alphabet Σ. It is based on three elementary edit operations: insertion, deletion and substitution of a symbol. Each operation has a specific cost, gathered in a nonnegative $(|\Sigma| + 1) \times (|\Sigma| + 1)$ matrix $\boldsymbol{C}$ (the additional row and column account for insertion and deletion costs respectively). A sequence of operations transforming a string x into a string x' is called an edit script. The edit distance between x and x' is defined as the cost of the cheapest edit script that turns x into x' and can be computed in $O(|\mathsf{x}| \cdot |\mathsf{x}'|)$ time by dynamic programming.[2]

The classic edit distance, known as the Levenshtein distance, uses a unit cost matrix and thus corresponds to the minimum number of operations turning one string into another. For instance, the Levenshtein distance between abb and aa is equal to 2, since turning abb into aa requires at least two operations (e.g., substitution of b with a and deletion of b). On the other hand, using the cost matrix given in Table 2.1, the edit distance between abb and aa is equal to 10 (deletion of a and two substitutions of b with a is the cheapest edit script). Using task-specific costs is a key to the performance of the edit distance in many applications. As we shall see in Chapter 7, there exist several methods to automatically learning these costs from data.

Sequence alignment Sequence alignment is a way of computing the similarity between two strings, mostly used in bioinformatics to identify regions of similarity in DNA or protein sequences [Mount, 2004]. It corresponds to the score of the best alignment, where an alignment score is based on the same elementary operations as the edit distance and on a score matrix for substitutions, but simply penalizes gaps in alignments using a linear or affine penalty function of the gap length instead of insertion and deletion costs. The most prominent sequence alignment

[2]Note that in the case of strings of equal length, the edit distance is upper bounded by the Hamming distance.

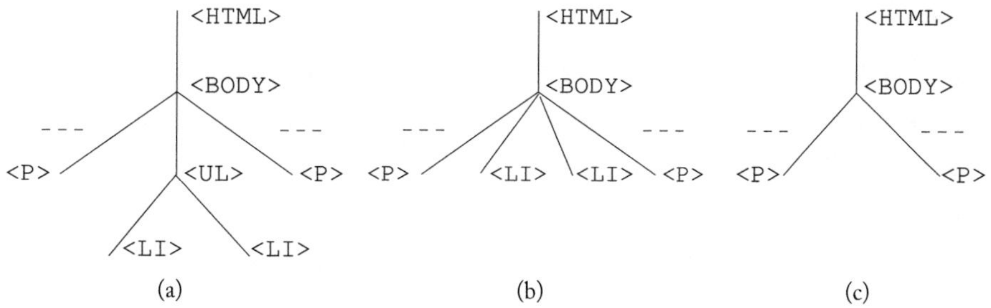

Figure 2.3: Strategies to delete a node within a tree: (a) original tree, (b) after deletion of the <UL> node as defined by Zhang & Shasha, and (c) after deletion of the <UL> node as defined by Selkow.

measures are the Needleman-Wunsch score [Needleman and Wunsch, 1970] for global alignments and the Smith-Waterman score [Smith and Waterman, 1981] for local alignments. They can be computed by dynamic programming.

Tree edit distance Because of the growing interest in applications that naturally involve tree-structured data (such as the secondary structure of RNA in biology, XML documents on the web or parse trees in natural language processing), several works have extended the string edit distance to trees, resorting to similar elementary edit operations [see Bille, 2005, for a survey on the matter]. The operations are the insertion, deletion and relabeling of a node.

There exist two main variants of the tree edit distance that differ in the way the deletion of a node is handled. In Zhang and Shasha [1989], when a node is deleted all its children are connected to its parent node. The best algorithms for computing this distance have an $O(n^3)$ worst-case complexity, where n is the number of nodes of the largest tree [see Pawlik and Augsten, 2011, for an empirical evaluation of several algorithms]. Another variant is due to Selkow [1977], where insertions and deletions are restricted to the leaves of the tree. Such a distance is relevant to specific applications. For instance, deleting a <UL> tag (i.e., a nonleaf node) of an unordered list in an HTML document would require the iterative deletion of the <LI> items (i.e., the subtree) first, which is a sensible thing to do in this context (see Figure 2.3). This version can be computed in quadratic time.[3]

Graph edit distance There also exist extensions of the edit distance to general graphs [Gao et al., 2010], but like many problems on graphs, computing a graph edit distance is NP-hard, making it impractical for real-world tasks.

Spectrum, subsequence and mismatch kernels These string kernels represent strings by fixed-length feature vectors and rely on explicit mapping functions ϕ. The spectrum kernel [Leslie et al.,

[3]Note that tree edit distance computations can be made significantly faster (especially for large trees) by exploiting lower bounds on the distance between two trees that are cheap to obtain [see for instance Yang et al., 2005].

2002a] maps each string to a vector of frequencies of all contiguous subsequences of length p and computes the inner product between these vectors. The subsequence kernel [Lodhi et al., 2002] and the mismatch kernel [Leslie et al., 2002b] extend the spectrum kernel to inexact subsequence matching: the former considers all (possibly noncontiguous) subsequences of length p while the latter allows a number of mismatches in the subsequences.

String edit kernels String edit kernels are derived from the string edit distance (or related measures). The classic edit kernel [Li and Jiang, 2004] has the following form:

$$K_{L\&J}(\mathsf{x}, \mathsf{x}') = e^{-t \cdot d_{lev}(\mathsf{x}, \mathsf{x}')},$$

where d_{lev} is the Levenshtein distance and $t > 0$ is a parameter. However, Cortes et al. [2004] have shown that this function is not PSD (and thus not a valid kernel) in general, so t must be carefully chosen.

Saigo et al. [2004] build a kernel from the sum of scores over all possible Smith-Waterman local alignments between two strings instead of the alignment of highest score only. They show that if the score matrix is PSD, then the kernel is valid in general.

2.3 METRICS IN MACHINE LEARNING AND DATA MINING

Metrics play an important role in many machine learning and data mining methods. We briefly review a few important examples. In all cases, the choice of the metric is key to the performance, providing a good motivation for metric learning.

Nearest neighbor methods Perhaps the most obvious use of metrics is in k-Nearest Neighbors (k-NN) for classification [Cover and Hart, 1967] and regression [Altman, 1992], where the prediction for a test instance corresponds to the majority class (or average value in the case of regression) among its k-nearest neighbors in the training set. The neighbors are determined based on a metric chosen by the user.

Kernel methods The key idea behind kernel methods [Schölkopf and Smola, 2001] is to use a kernel function to implicitly map data to a high-dimensional nonlinear space while allowing cheap computations of inner products in that space through the kernel. This is known as the "kernel trick" and has been used to obtain powerful and efficient nonlinear versions of various linear methods for classification, regression, dimensionality reduction, etc. Prominent examples include Support Vector Machines (SVM) classification [Cortes and Vapnik, 1995], where a large-margin linear classifier is learned implicitly in kernel space, the nonlinear regression algorithm kernel ridge [Saunders et al., 1998] as well as kernel Principal Component Analysis (PCA) [Schölkopf et al., 1998], a kernelized version of the popular dimensionality reduction method [Pearson, 1901].

Clustering Clustering [Xu and Wunsch, 2008] is the task of grouping objects such that those in the same group are more similar to each other than those in different groups. For instance, given a number of clusters, algorithms such as K-Means [Lloyd, 1982] and K-Medoids [Kaufman and Rousseeuw, 1987] essentially aim at minimizing the distance between points assigned to a cluster and the center of that cluster.

Information retrieval A key task in information retrieval [Manning et al., 2008] is to retrieve information from a database that is most relevant to a given query. For instance, internet search engines aim at retrieving webpages that are relevant to a user's text query. Another example is image search, where the goal is to find the images in the database that are the most similar to a query image. In all cases, one can use a similarity measure to rank the documents by relevance [see for instance Baeza-Yates and Ribeiro-Neto, 1999, Salton et al., 1975, Sivic and Zisserman, 2009].

Data visualization Data in high dimensions are difficult to visualize. One way to discover interesting patterns in such data is to use a metric to plot data points based on their distance to one another [see for instance Bertini et al., 2011, van der Maaten and Hinton, 2008, Venna et al., 2010].

Structured data Metrics are especially convenient when dealing with structured data because they can be used as a proxy to manipulate these complex objects. Indeed, any algorithm that only accesses data through a metric (such as those presented above) may be used as if the data consisted of feature vectors. For instance, the development of structured kernels [Gärtner, 2003] allowed kernel methods to be readily applicable to various types of structured objects.

CHAPTER 3

Properties of Metric Learning Algorithms

Before delving into specific methods, this chapter introduces five key properties of metric learning algorithms: type of metric, form of supervision, scalability, optimality guarantees and ability to perform dimensionality reduction (Figure 3.1). When deciding which method to apply, emphasis should be placed on these properties, depending on the characteristics of the problem at hand. They provide the basis for a taxonomy of all the algorithms covered in this book that we will use to summarize the literature in Chapter 10.

Learning paradigm We will consider three learning paradigms:

- *Fully supervised*: the metric learning algorithm has access to a set of labeled training instances $\{z_i = (x_i, y_i)\}_{i=1}^n$, where each training example $z_i \in \mathcal{Z} = \mathcal{X} \times \mathcal{Y}$ is composed of an instance $x_i \in \mathcal{X}$ and a label (or class) $y_i \in \mathcal{Y}$. $\mathcal{Y}$ is a discrete and finite set of $|\mathcal{Y}|$ labels (unless stated otherwise). In practice, the label information is often used to generate specific sets of pair/triplet constraints $\mathcal{S}, \mathcal{D}, \mathcal{R}$, for instance based on a notion of neighborhood.

- *Weakly supervised*: the algorithm has no access to the labels of training instances: it is only provided with side information in the form of sets of constraints $\mathcal{S}$, $\mathcal{D}$ and $\mathcal{R}$. This setting is common in a variety of applications where labeled data is costly to obtain while such side information is cheap: examples include users' implicit feedback (e.g., clicks on search engine results), citations among articles or links in a network. This can be seen as having label information only at the pair/triplet level.

- *Semi-supervised*: besides the (full or weak) supervision, the algorithm has access to a (typically large) sample of unlabeled instances for which no side information is available. Such data can be useful to avoid overfitting when the labeled data or side information are scarce.

Form of metric The form of the metric is a key choice in metric learning and belongs to three main cases:

- *Linear metrics*, such as the Mahalanobis distance. Their expressive power is limited but they are typically easy to optimize (formulations are typically convex with global optimality of the solution) and suffer little overfitting.

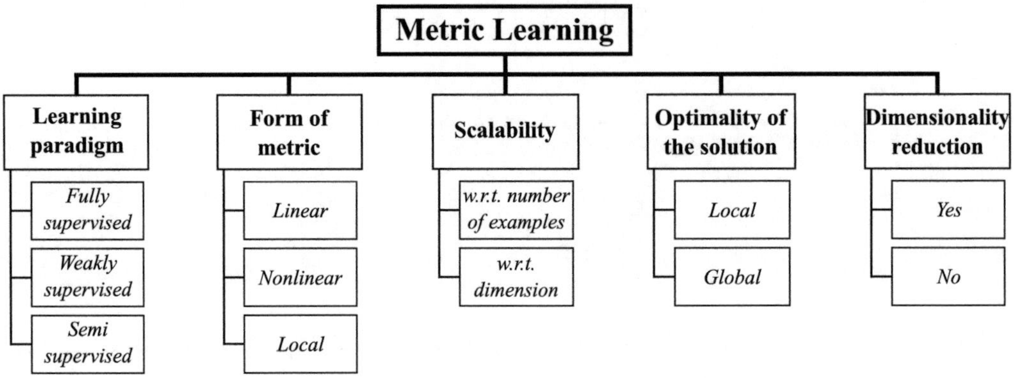

Figure 3.1: Five key properties of metric learning algorithms.

- *Nonlinear metrics*, such as the χ^2 histogram distance. They often give rise to nonconvex formulations (subject to local optimality) and may overfit, but they can capture nonlinear variations in the data.
- *Local metrics*, where multiple (typically linear) metrics are learned in several regions of the feature space to better deal with complex problems, such as heterogeneous data. They are however prone to overfitting since the number of parameters to learn can be very large.

Scalability With the fast increasing amounts of data available for training in many domains, the problem of scalability arises in all areas of machine learning. First, it is desirable for a metric learning algorithm to scale well with the number of training examples/constraints. Second, it should scale reasonably well with the dimensionality d of the data, which is often difficult since metric learning is often phrased as learning a $d \times d$ matrix.

Optimality of the solution This property refers to the ability of the algorithm to find the parameters that optimize the objective function of interest. Ideally, the solution is guaranteed to be the *global optimum*, which is essentially the case for convex formulations of metric learning. On the other hand, for nonconvex formulations, the solution may only be a *local optimum*.

Dimensionality reduction Metric learning is sometimes formulated as finding a projection of the data into a new feature space. An interesting byproduct in this case is to learn a metric that induces a low-dimensional space, resulting in faster distance computations as well as more compact representations of data.

CHAPTER 4

Linear Metric Learning

This chapter reviews the literature on learning a linear metric, i.e., computed as a linear function of its inputs. Such metrics are amenable to efficient learning thanks to their simple form. The chapter is organized as follows. In the first part, we focus on Malahanobis distance learning (Section 4.1), where the learned metric satisfies the distance axioms. Then, motivated by psychological evidence and computational benefits, the second part is devoted to learning more flexible linear similarity functions (Section 4.2). Finally, Section 4.3 discusses how to scale-up these methods to large amounts of training data (both in number of samples and number of features).

4.1 MAHALANOBIS DISTANCE LEARNING

This section deals with Malahanobis distance learning, which has attracted a lot of interest due to its nice interpretation in terms of a linear projection (see Section 2.2.1). Recall that a Mahalanobis distance has the form

$$d_{\boldsymbol{M}}(\boldsymbol{x}, \boldsymbol{x}') = \sqrt{(\boldsymbol{x} - \boldsymbol{x}')^T \boldsymbol{M} (\boldsymbol{x} - \boldsymbol{x}')}$$

and is parameterized by $\boldsymbol{M} \in \mathbb{S}_+^d$, where $\mathbb{S}_+^d$ is the cone of PSD $d \times d$ real-valued matrices. In practice, a Mahalanobis distance is learned in its more convenient squared form:

$$d_{\boldsymbol{M}}^2(\boldsymbol{x}, \boldsymbol{x}') = (\boldsymbol{x} - \boldsymbol{x}')^T \boldsymbol{M} (\boldsymbol{x} - \boldsymbol{x}').$$

The goal of Mahalanobis distance learning is to learn $\boldsymbol{M} \in \mathbb{S}_+^d$. Maintaining the PSD constraint in an efficient way during the optimization process is a key issue that attracted significant interest, as we shall see later.

The rest of this section is a review of the main Mahalanobis distance learning methods of the literature. We first present in Section 4.1.1 three early approaches of historical and practical importance. In Section 4.1.2, we focus on regularization-based methods for Mahalanobis distance learning.

4.1.1 EARLY APPROACHES

This section presents some early influential work that largely contributed to the popularity of metric learning and paved the way for subsequent research in the field.

MMC (Xing et al.) The seminal work of Xing et al. [2002] is the first Mahalanobis distance learning method.[1] The goal is to learn a distance function for use in clustering based on two

[1]Source code available at: `http://www.cs.cmu.edu/~epxing/papers/`.

sets of pairs of similar and dissimilar examples ($\mathcal{S}$ and $\mathcal{D}$ respectively). The idea builds on the between- and within-cluster variation: maximize the sum of distances between dissimilar points while keeping the sum of distances between similar examples small. This can be formulated as follows:

$$\begin{aligned} \max_{\boldsymbol{M} \in \mathbb{S}_+^d} \quad & \sum_{(\boldsymbol{x}_i, \boldsymbol{x}_j) \in \mathcal{D}} d_{\boldsymbol{M}}(\boldsymbol{x}_i, \boldsymbol{x}_j) \\ \text{s.t.} \quad & \sum_{(\boldsymbol{x}_i, \boldsymbol{x}_j) \in \mathcal{S}} d_{\boldsymbol{M}}^2(\boldsymbol{x}_i, \boldsymbol{x}_j) \leq 1. \end{aligned} \tag{4.1}$$

The problem (4.1) is convex and always feasible (setting for instance $\boldsymbol{M} = \mathbf{0}$). The authors propose to solve it using a projected gradient descent algorithm. At each iteration, a gradient ascent step is taken on the objective function, followed by a projection onto the sets $C_1 = \{\boldsymbol{M} : \sum_{(\boldsymbol{x}_i, \boldsymbol{x}_j) \in \mathcal{S}} d_{\boldsymbol{M}}^2(\boldsymbol{x}_i, \boldsymbol{x}_j) \leq 1\}$ and $C_2 = \{\boldsymbol{M} : \boldsymbol{M} \in \mathbb{S}_+^d\}$. The first projection step can be done in $O(d^2)$ time by solving a system of linear equations. The second one is a projection onto the PSD cone, which is done by setting the negative eigenvalues of $\boldsymbol{M}$ to zero. One thus needs to compute the eigenvalue decomposition of $\boldsymbol{M}$, which requires $O(d^3)$ time. These steps are repeated until convergence. Because of the PSD projection cost, the approach becomes quite inefficient when d is larger than a few hundreds.

Experiments on standard low-dimensional datasets show that the learned distance consistently improves K-Means clustering performance. MMC can also be combined with clustering algorithms that incorporate distance constraints such as Constrained K-Means [Wagstaff et al., 2001].

Recent work [Cao et al., 2012b, Ying and Li, 2012] revisited MMC by casting it as an eigenvalue optimization problem and showed that it can be solved by an algorithm that only requires the computation of the largest eigenvalue of $\boldsymbol{M}$ at each iteration. This reduces the time complexity in d of an iteration to $O(d^2)$.

NCA (Goldberger et al.) The idea of neighborhood Component Analysis[2] (NCA), introduced by Goldberger et al. [2004], is to optimize the expected leave-one-out error of a stochastic nearest neighbor classifier in the projection space induced by $d_{\boldsymbol{M}}$. Given a training sample $\{\boldsymbol{x}_i, y_i\}_{i=1}^n$, they define the probability that $\boldsymbol{x}_i$ is the neighbor of $\boldsymbol{x}_j$ by

$$p_{ij} = \frac{\exp(-\|\boldsymbol{L}\boldsymbol{x}_i - \boldsymbol{L}\boldsymbol{x}_j\|_2^2)}{\sum_{l \neq i} \exp(-\|\boldsymbol{L}\boldsymbol{x}_i - \boldsymbol{L}\boldsymbol{x}_l\|_2^2)}, \quad p_{ii} = 0, \tag{4.2}$$

where $\boldsymbol{M} = \boldsymbol{L}^T\boldsymbol{L}$. Then, the probability that $\boldsymbol{x}_i$ is correctly classified is:

$$p_i = \sum_{j : y_j = y_i} p_{ij}.$$

[2]Source code available at: `http://www.ics.uci.edu/~fowlkes/software/nca/`.

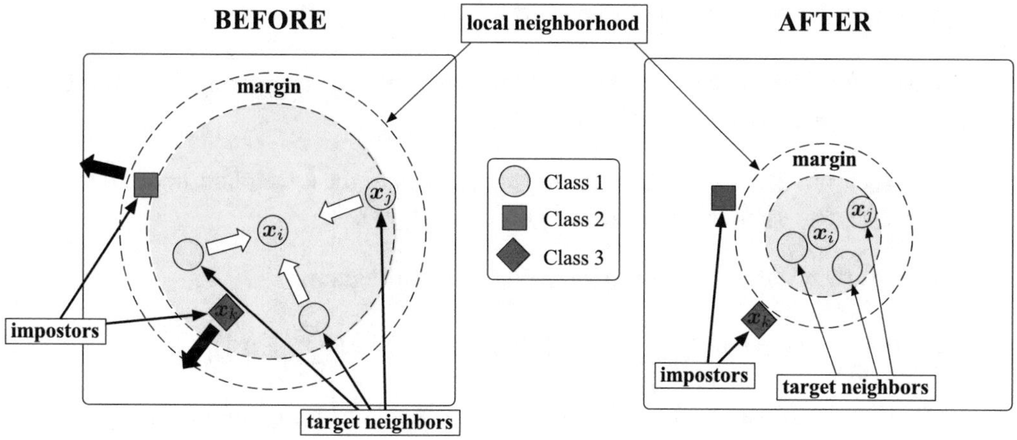

Figure 4.1: Illustration of the neighborhood-based constraints used by LMNN (adapted from Weinberger and Saul [2009]).

The distance is learned by maximizing the sum of these probabilities:

$$\max_{\boldsymbol{L}} \sum_{i=1}^{n} p_i. \tag{4.3}$$

Problem (4.3) is solved using gradient descent, but the formulation is nonconvex and thus the algorithm can get stuck in bad local optima. Nevertheless, experiments show that the learned metric outperforms the Euclidean distance in k-NN classification and can be used to reduce dimensionality by choosing a rectangular $\boldsymbol{L}$. Note that computing the normalization in (4.2) for all training instances requires $O(n^2)$ time, preventing the application of NCA to large datasets.

In follow-up work, Globerson and Roweis [2005] proposed an alternative formulation of NCA based on minimizing a KL divergence between p_{ij} and an ideal distribution, which can be seen as attempting to collapse each class to a single point. The resulting problem is convex, but requires PSD projections. Finally, note that Tarlow et al. [2013] generalized NCA to k-NN with $k > 1$.

LMNN (Weinberger et al.) Large Margin Nearest Neighbors[3] (LMNN), introduced by Weinberger et al. [2008, 2009, 2005], is probably the most widely known metric learning algorithm and served as the basis for many extensions (some of which are described in later sections of this book).

The main reason for its popularity lies in the way the training constraints are defined. Motivated by k-NN classification, they are constructed from a labeled training set $\{\boldsymbol{x}_i, y_i\}_{i=1}^n$ in a *local*

[3]Source code available at: `http://www.cse.wustl.edu/~kilian/code/code.html`.

way. Specifically, for each training instance, the k nearest neighbors of the same class (the "target neighbors") should be closer than instances of other classes (the "impostors")—see Figure 4.1 for an illustration. The Euclidean distance is used to determine target neighbors and impostors. Formally, the constraints are defined as follows:

$$\begin{aligned} \mathcal{S}_{lmnn} &= \{(\boldsymbol{x}_i, \boldsymbol{x}_j) : y_i = y_j \text{ and } \boldsymbol{x}_j \text{ belongs to the } k\text{-neighborhood of } \boldsymbol{x}_i\}, \\ \mathcal{R}_{lmnn} &= \{(\boldsymbol{x}_i, \boldsymbol{x}_j, \boldsymbol{x}_k) : (\boldsymbol{x}_i, \boldsymbol{x}_j) \in \mathcal{S}, y_i \neq y_k\}. \end{aligned}$$

The distance is then learned using the following convex program:

$$\begin{aligned} \min_{\boldsymbol{M} \in \mathbb{S}_+^d, \xi \geq 0} \quad & (1-\mu) \sum_{(\boldsymbol{x}_i, \boldsymbol{x}_j) \in \mathcal{S}_{lmnn}} d_{\boldsymbol{M}}^2(\boldsymbol{x}_i, \boldsymbol{x}_j) \quad + \quad \mu \sum_{i,j,k} \xi_{ijk} \\ \text{s.t.} \quad & d_{\boldsymbol{M}}^2(\boldsymbol{x}_i, \boldsymbol{x}_k) - d_{\boldsymbol{M}}^2(\boldsymbol{x}_i, \boldsymbol{x}_j) \geq 1 - \xi_{ijk} \quad \forall (\boldsymbol{x}_i, \boldsymbol{x}_j, \boldsymbol{x}_k) \in \mathcal{R}_{lmnn}, \end{aligned} \tag{4.4}$$

where $\mu \in [0, 1]$ controls the trade-off between pulling target neighbors closer together and pushing away impostors.

Note that the number of constraints in Problem (4.4) is in the order of kn^2 and thus grows quickly with n. Fortunately, because of the way $\mathcal{R}_{lmnn}$ is built, many constraints are almost trivially satisfied. The subgradient descent algorithm proposed by the authors takes advantage of this by maintaining a working set of violated contraints through careful book-keeping. Another popular workaround in subsequent metric learning literature is to consider only the closest impostors instead of all differently labeled points [see for instance Shen et al., 2009, Shi et al., 2014, Wang et al., 2012c]. We also point out that alternative algorithms for solving the LMNN problem have been proposed [Nguyen and Guo, 2008, Park et al., 2011, Torresani and Lee, 2006].

LMNN achieves good practical performance in k-NN classification: it is generally on par with NCA but is easier to scale to problems with larger n. Unlike NCA, it is however dependent on the Euclidean distance's ability to select relevant target neighbors. To conclude, we note that a relation between LMNN and Support Vector Machines was pointed out by Do et al. [2012].

4.1.2 REGULARIZED APPROACHES

A central problem in machine learning is to achieve a good "bias-variance" trade-off. On the one hand, the model should be complex enough to perform well on the training data (low bias). On the other hand, it should be simple enough so that it generalizes well to unseen data (low variance). A popular way to implement this trade-off to select a model with good bias-variance trade-off is through the use of regularization. In the context of metric learning, this can be formulated as the following general form:

$$\min_{\boldsymbol{M}} \quad \ell(\boldsymbol{M}, \mathcal{S}, \mathcal{D}, \mathcal{R}) \quad + \quad \lambda R(\boldsymbol{M})$$

where $\ell(\boldsymbol{M}, \mathcal{S}, \mathcal{D}, \mathcal{R})$ is a loss function that incurs a penalty when training constraints are violated, $R(\boldsymbol{M})$ is a regularizer that penalizes the complexity of the metric and $\lambda \geq 0$ is the parameter controlling the trade-off (typically tuned on some validation set).

In recent years, there has been a lot of interest in sparsity-inducing norms for the regularization of machine learning models. These regularizers favor models that are sparse in some way, e.g., that have many zero coefficients. A good account of the properties of these regularizers and the optimization algorithms to solve the resulting problems can be found in [Bach et al., 2012].

The approaches presented in Section 4.1.1 do not explicitly regularize the metric. In this section, we review a few important methods that rely on regularization to prevent from overfitting and, in some cases, to make the algorithm more efficient or to favor to metrics with interesting properties (e.g., sparsity).

S&J (Schultz & Joachims) The method of Schultz and Joachims [2003] proposes to learn a metric for classification and is inspired by Support Vector Machines [Cortes and Vapnik, 1995]. The "hard-margin" version of the approach can be formulated as follows:

$$\begin{aligned} \min_{\boldsymbol{M} \in \mathbb{S}_+^d} \quad & \|\boldsymbol{M}\|_{\mathcal{F}}^2 \\ \text{s.t.} \quad & d_{\boldsymbol{M}}^2(\boldsymbol{x}_i, \boldsymbol{x}_k) - d_{\boldsymbol{M}}^2(\boldsymbol{x}_i, \boldsymbol{x}_j) \geq 1 \quad \forall (\boldsymbol{x}_i, \boldsymbol{x}_j, \boldsymbol{x}_k) \in \mathcal{R}, \end{aligned} \tag{4.5}$$

where $\|\boldsymbol{M}\|_{\mathcal{F}}^2 = \sum_{i,j} M_{ij}^2$ is the squared Frobenius norm of $\boldsymbol{M}$, a simple regularizer to avoid overfitting (like the L_2-norm used in SVM). Notice that Problem (4.5) requires that each triplet constraint is satisfied with margin 1. Because this may be infeasible, we instead work with a more flexible "soft-margin" version:

$$\begin{aligned} \min_{\boldsymbol{M} \in \mathbb{S}_+^d, \xi \geq 0} \quad & \sum_{i,j,k} \xi_{ijk} \quad + \quad \lambda \|\boldsymbol{M}\|_{\mathcal{F}}^2 \\ \text{s.t.} \quad & d_{\boldsymbol{M}}^2(\boldsymbol{x}_i, \boldsymbol{x}_k) - d_{\boldsymbol{M}}^2(\boldsymbol{x}_i, \boldsymbol{x}_j) \geq 1 - \xi_{ijk} \quad \forall (\boldsymbol{x}_i, \boldsymbol{x}_j, \boldsymbol{x}_k) \in \mathcal{R}, \end{aligned} \tag{4.6}$$

where the ξ_{ijk}'s are nonnegative "slack" variables to penalize the objective function whenever a constraint is not satisfied with at least margin 1,[4] and $\lambda \geq 0$ is the parameter adjusting the trade-off between regularization and constraint satisfaction. Note that Problem (4.6) is convex.

Many efficient SVM solvers have been developed. To be able to use them for Problem (4.6), the authors parameterize $\boldsymbol{M} = \boldsymbol{A}^T \boldsymbol{W} \boldsymbol{A}$, where $\boldsymbol{A}$ is a real matrix fixed in advance, and $\boldsymbol{W}$ is a diagonal matrix with nonnegative entries. We get:

$$d_{\boldsymbol{M}}^2(\boldsymbol{x}_i, \boldsymbol{x}_j) = (\boldsymbol{A}\boldsymbol{x}_i - \boldsymbol{A}\boldsymbol{x}_j)^T \boldsymbol{W} (\boldsymbol{A}\boldsymbol{x}_i - \boldsymbol{A}\boldsymbol{x}_j).$$

In other words, $\boldsymbol{A}$ linearly projects the data into a new space and $\boldsymbol{W}$ weights the resulting features. By definition, $\boldsymbol{M}$ is PSD and one can simply optimize over the diagonal matrix $\boldsymbol{W}$, avoiding the need for projections onto the PSD cone. This method can thus be applied to relatively high-dimensional problems. The downside is that it is less general than full Mahalanobis distance learning, as one only learns a weighting of the features. Furthermore, $\boldsymbol{A}$ must be chosen by the user and this choice can greatly affect the classification performance.

[4]Throughout the book, we will consistently use the symbol ξ to denote slack variables.

ITML (Davis et al.) Information-Theoretic Metric Learning[5] (ITML), introduced by Davis et al. [2007], proposes the use of LogDet regularization for metric learning. The idea is to regularize the matrix $\boldsymbol{M}$ to remain as close as possible to a good prior matrix $\boldsymbol{M}_0$, such as the identity matrix (Euclidean distance) or the inverse covariance.

The authors propose to measure the "closeness" between two $d \times d$ positive definite matrices $\boldsymbol{M} = \boldsymbol{V}\boldsymbol{\Lambda}\boldsymbol{V}^T$ and $\boldsymbol{M}_0 = \boldsymbol{U}\boldsymbol{\Theta}\boldsymbol{U}^T$ using a Bregman divergence called the LogDet:

$$\begin{aligned} D_{ld}(\boldsymbol{M}, \boldsymbol{M}_0) &= \operatorname{tr}(\boldsymbol{M}\boldsymbol{M}_0^{-1}) - \log\det(\boldsymbol{M}\boldsymbol{M}_0^{-1}) - d \\ &= \sum_{i,j} \frac{\lambda_i}{\theta_j}(\boldsymbol{v}_i^T \boldsymbol{u}_i)^2 - \sum_i \log\left(\frac{\lambda_i}{\theta_i}\right) - d. \end{aligned} \tag{4.7}$$

It can be shown that $D_{ld}(\boldsymbol{M}, \boldsymbol{M}_0)$ is equivalent to minimizing the KL divergence between the multivariate Gaussian distributions parameterized by $\boldsymbol{M}$ and $\boldsymbol{M}_0$. Furthermore, we can see from (4.7) that $D_{ld}(\boldsymbol{M}, \boldsymbol{M}_0)$ is finite if and only if the range of $\boldsymbol{M}$ is equal to the range of $\boldsymbol{M}_0$. Therefore minimizing $D_{ld}(\boldsymbol{M}, \boldsymbol{M}_0)$ provides an automatic way to ensure that $\boldsymbol{M}$ is positive definite and has the same rank as $\boldsymbol{M}_0$.

Building on these observations, ITML is formulated as a trade-off between minimizing the LogDet divergence and satisfying pairwise constraints:

$$\begin{aligned} \min_{\boldsymbol{M} \in \mathbb{S}_+^d} \quad & \sum_{i,j} \xi_{ij} \quad + \quad \lambda D_{ld}(\boldsymbol{M}, \boldsymbol{M}_0) \\ \text{s.t.} \quad & d_{\boldsymbol{M}}^2(\boldsymbol{x}_i, \boldsymbol{x}_j) \leq u + \xi_{ij} \quad \forall (\boldsymbol{x}_i, \boldsymbol{x}_j) \in \mathcal{S} \\ & d_{\boldsymbol{M}}^2(\boldsymbol{x}_i, \boldsymbol{x}_j) \geq v - \xi_{ij} \quad \forall (\boldsymbol{x}_i, \boldsymbol{x}_j) \in \mathcal{D}, \end{aligned} \tag{4.8}$$

where $u, v, \lambda \geq 0$ are parameters and $\boldsymbol{M}_0$ is a fixed positive definite matrix. Problem (4.8) has soft constraints to keep distances between similar examples smaller than u and those between dissimilar examples larger than v. The LogDet divergence is convex in $\boldsymbol{M}$ (because the determinant of a positive definite matrix is log-concave) and thus the problem is convex.

The proposed algorithm is based on repeatedly performing Bregman projections onto a single constraint [Kulis et al., 2006]. Namely, at iteration t, we pick a constraint $(\boldsymbol{x}_i, \boldsymbol{x}_j)$ in $\mathcal{S}$ or $\mathcal{D}$ and update using

$$\boldsymbol{M}^{t+1} = \boldsymbol{M}^t + \beta \boldsymbol{M}^t(\boldsymbol{x}_i - \boldsymbol{x}_j)(\boldsymbol{x}_i - \boldsymbol{x}_j)^T \boldsymbol{M}^t, \tag{4.9}$$

where β is the projection parameter that essentially depends on λ, the constraint type and its current violation. Computing (4.9) is done in $O(d^2)$ time and no eigen-decomposition is needed. Experiments on standard datasets show that ITML with $\boldsymbol{M}_0 = \boldsymbol{I}$ is on par with LMNN in terms of k-NN classification performance. A limitation of ITML is that the choice of $\boldsymbol{M}_0$ may have an important influence on the performance of the learned distance.

Several subsequent metric learning methods adopted LogDet divergence for regularization [e.g., Jain et al., 2008, Qi et al., 2009]. Finally, note that Kulis et al. [2009] have shown how hashing can be used together with ITML to achieve fast similarity search.

[5]Source code available at: `http://www.cs.utexas.edu/~pjain/itml/`.

SML (Ying et al.) SML[6] [Ying et al., 2009], for Sparse Metric Learning, is a distance learning approach that regularizes $\boldsymbol{M}$ with the mixed $L_{2,1}$ norm defined as

$$\|\boldsymbol{M}\|_{2,1} = \sum_{i=1}^{d} \|\boldsymbol{M}_i\|_2.$$

In other words, it is the L_1 norm of the L_2 norms of the columns. Unlike the L_1 norm used in previous work [Rosales and Fung, 2006] which favors sparsity at the entry level, the $L_{2,1}$ norm tends to zero out entire columns of $\boldsymbol{M}$, and can thus perform feature selection.

Formally, the authors set $\boldsymbol{M} = \boldsymbol{U}^T \boldsymbol{W} \boldsymbol{U}$, where $\boldsymbol{U} \in \mathbb{O}^d$ (the set of $d \times d$ orthonormal matrices) and $\boldsymbol{W} \in \mathbb{S}_+^d$, and solve the following problem:

$$\begin{aligned} \min_{\boldsymbol{U} \in \mathbb{O}^d, \boldsymbol{W} \in \mathbb{S}_+^d} \quad & \|\boldsymbol{W}\|_{2,1} \quad + \quad \lambda \sum_{i,j,k} \xi_{ijk} \\ \text{s.t.} \quad & d_M^2(\boldsymbol{x}_i, \boldsymbol{x}_k) - d_M^2(\boldsymbol{x}_i, \boldsymbol{x}_j) \geq 1 - \xi_{ijk} \quad \forall (\boldsymbol{x}_i, \boldsymbol{x}_j, \boldsymbol{x}_k) \in \mathcal{R}, \end{aligned} \tag{4.10}$$

Problem (4.10) is reformulated as a min-max problem and solved using smooth optimization [Nesterov, 2005]. The proposed algorithm has an iteration cost with $O(d^3)$ complexity, but more recent optimization techniques such as proximal methods could potentially reduce this cost [see for instance Bach et al., 2012].

SML achieves slightly better k-NN performance than LMNN or ITML on some UCI datasets. This is largely due to its ability to ignore irrelevant feature dimensions provided by the $L_{2,1}$ norm regularization.

BoostMetric (Shen et al.) BoostMetric[7] [Shen et al., 2009, 2012] adapts to Mahalanobis distance learning the ideas of boosting, where a good hypothesis is obtained through a weighted combination of so-called "weak learners" [Schapire and Freund, 2012]. Inspired from Adaboost [Freund and Schapire, 1995], they propose a formulation based on the exponential loss:

$$\min_{\boldsymbol{M} \in \mathbb{S}_+^d} \quad \log \sum_{r=1}^{|\mathcal{R}|} \exp(-\rho_r) \quad + \quad \lambda \operatorname{tr}(\boldsymbol{M}), \tag{4.11}$$

where for ease of notation, each constraint in $\mathcal{R}$ is represented as $\rho_r = \langle \boldsymbol{A}_r, \boldsymbol{M} \rangle$ with

$$\boldsymbol{A}_r = (\boldsymbol{x}_i - \boldsymbol{x}_k)(\boldsymbol{x}_i - \boldsymbol{x}_k)^T - (\boldsymbol{x}_i - \boldsymbol{x}_j)(\boldsymbol{x}_i - \boldsymbol{x}_j)^T.$$

The trace of $\boldsymbol{M}$ is used as a regularizer. In the case of PSD matrices, it is equivalent to the trace norm (also known as the nuclear norm) which favors low-rank solutions [see Kunapuli and Shavlik, 2012, for another approach based on the trace norm].

[6] Source code is not available but is listed as "coming soon" by the authors. Check: `http://www.enm.bris.ac.uk/staff/xyy/software.html`

[7] Source code available at: `http://code.google.com/p/boosting/`.

BoostMetric is based on the property that any PSD matrix can be decomposed into a non-negative linear combination of trace-one rank-one matrices:

$$M = \sum_{j=1}^{J} w_j Z_j \quad \text{where } \forall j : w_j \geq 0, Z_j = v_j v_j^T, v_j \in \mathbb{R}^d, \|v_j\|_2 = 1.$$

Instead of solving (4.11), one can solve the dual problem:

$$\begin{aligned} \max_{u \in \mathbb{R}^{|\mathcal{R}|}} \quad & -\sum_{r=1}^{|\mathcal{R}|} u_r \log u_r \\ \text{s.t.} \quad & \sum_{r=1}^{|\mathcal{R}|} u_r = 1, \quad u_r \geq 0 \\ & \sum_{r=1}^{|\mathcal{R}|} u_r \langle A_r, Z_j \rangle \leq \lambda, \quad 1 \leq j \leq J. \end{aligned} \tag{4.12}$$

Unfortunately, we do not know which Z_j's have nonzero weight and the set of candidates is infinite. To go around this problem, the authors show that one can solve (4.12) by an algorithm similar to Adaboost, where the weak learners are the trace-one rank-one matrices. At each iteration, the algorithm incorporates a new learner Z to the solution by solving

$$Z = \arg\max_{Z'} \left\{ \sum_{r=1}^{|\mathcal{R}|} u_r \langle A_r, Z' \rangle \right\},$$

where u_r is proportional to the current violation of constraint r. Z can be found by computing the largest eigenvector of $\sum_{r=1}^{|\mathcal{R}|} u_r A_r$, which requires $O(d^2)$ time.

In practice, BoostMetric achieves competitive performance but typically requires a very large number of iterations for high-dimensional datasets. In subsequent work, Bi et al. [2011] further improve the scalability of the approach, while Liu and Vemuri [2012] introduce regularization on the weights as well as a term to reduce redundancy among weak learners.

4.2 LINEAR SIMILARITY LEARNING

The work reviewed so far has focused on learning a Mahalanobis distance, which satisfies the distance axioms. While these axioms are intuitive and can be useful in practice, for instance to speed up neighbor search [Uhlmann, 1991], they impose strict constraints that may be too rigid. In fact, the question of whether human similarity judgments satisfy the distance axioms has been extensively debated in cognitive psychology, with some experiments challenging all of them, including symmetry and triangle inequality [Ashby and Perrin, 1988, Rosch, 1975, Tversky, 1977, Tversky and Gati, 1982]. Figure 4.2 illustrates this in the case of visual similarity. These studies

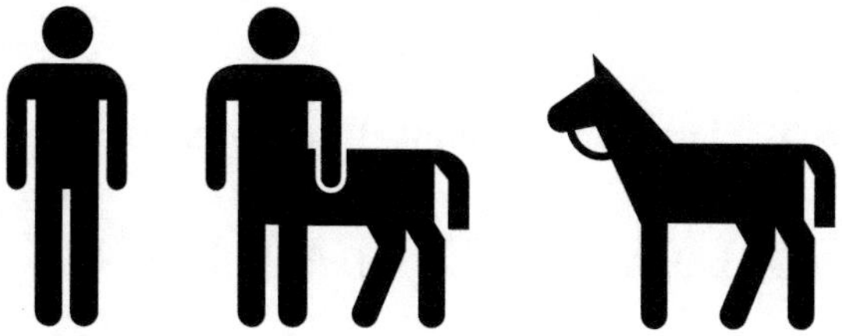

Figure 4.2: Visual similarity judgments can violate the triangle inequality. In this example, a man and a horse are both similar to a centaur, but not to one another.

suggest that more flexible forms of similarity should be used when the goal is to approximate perceptual similarity. There can also be a computational benefit: for instance, lifting the PSD constraint typically leads to more scalable algorithms. In this section, we summarize the existing work on learning linear similarity functions that do not satisfy the distance axioms.

Qamar et al. [2008] propose to learn a similarity function of the following form:

$$S_M(\boldsymbol{x}, \boldsymbol{x}') = \frac{\boldsymbol{x}^T \boldsymbol{M} \boldsymbol{x}'}{N(\boldsymbol{x}, \boldsymbol{x}')},$$

where $\boldsymbol{M} \in \mathbb{R}^{d \times d}$ is not required to be PSD or symmetric, and $N(\boldsymbol{x}, \boldsymbol{x}')$ is a normalization term which depends on $\boldsymbol{x}$ and $\boldsymbol{x}'$. This similarity function can be seen as a generalization of the cosine similarity, widely used in text and image retrieval [see for instance Baeza-Yates and Ribeiro-Neto, 1999, Sivic and Zisserman, 2009]. To learn $\boldsymbol{M}$, the authors build on the notion of "target neighbors" introduced in LMNN. In subsequent work, Qamar and Gaussier [2009] propose a convex formulation to learn generalized cosine similarities of the form

$$S_M(\boldsymbol{x}, \boldsymbol{x}') = \frac{\boldsymbol{x}^T \boldsymbol{M} \boldsymbol{x}'}{\sqrt{\boldsymbol{x}^T \boldsymbol{M} \boldsymbol{x}} \sqrt{\boldsymbol{x}'^T \boldsymbol{M} \boldsymbol{x}'}},$$

where $\boldsymbol{M} \in \mathbb{S}_+^d$. This corresponds to a cosine similarity in the projection space induced by $\boldsymbol{M}$, but is more costly to optimize due to the PSD constraint. Nguyen and Bai [2010] optimize the same form of similarity based on a nonconvex formulation.

Other work has focused on the even simpler bilinear similarity:

$$S_M(\boldsymbol{x}, \boldsymbol{x}') = \boldsymbol{x}^T \boldsymbol{M} \boldsymbol{x}',$$

where $\boldsymbol{M} \in \mathbb{R}^{d \times d}$. OASIS [Chechik et al., 2009, 2010] deals with large-scale problems in image classification (refer to Section 4.3.1 for details). SLLC [Bellet et al., 2012b] takes an original angle by learning a similarity for sparse linear classification with generalization guarantees (the approach is presented and analyzed in Section 8.3.1). Cheng [2013] learn the similarity for pair matching (predicting whether two examples are similar) when the examples composing the pair may have different dimensionality. This is a relevant setting for matching instances from different domains, such as images with different resolutions, or queries and documents. They learn a rectangular matrix with a rank constraint based on techniques from optimization over Riemannian manifolds.

4.3 LARGE-SCALE METRIC LEARNING

In recent years, the volume of data available for training machine learning models has been increasing dramatically, both in the number of samples n and the number of features d. Dealing with these massive datasets imposes severe constraints on the time and memory complexity of the algorithms, introducing different trade-offs in learning [Bottou and Bousquet, 2007]. In this section, we discuss several ways to scale-up metric learning. These techniques could also be applied to some of the metric learning methods that we will present in Chapter 5 and Chapter 6.

4.3.1 LARGE n: ONLINE, STOCHASTIC AND DISTRIBUTED OPTIMIZATION

We start with the case where the number of training samples n (and therefore the number of training constraints) is very large, say in the millions. In this case, careful subsampling of the constraints may be sufficient to achieve near-optimal performance [see for instance Clémençon et al., 2015]. Otherwise, techniques such as online learning and stochastic optimization can be used to derive algorithms whose iteration cost is independent of n.

Online Approaches

In online learning [Cesa-Bianchi and Lugosi, 2006, Littlestone, 1988], the algorithm receives training instances one at a time and updates at each step the current hypothesis. Although the performance of online algorithms is typically inferior to batch algorithms, they are very useful to tackle large-scale problems that batch methods fail to address due to time and space complexity issues. Online learning methods often come with regret bounds, stating that the accumulated loss suffered along the way is not much worse than that of the best hypothesis chosen in hindsight:

$$\sum_{t=1}^{T} \ell(h_t, z_t) - \sum_{t=1}^{T} \ell(h^*, z_t) \leq f(T),$$

where T is the number of steps, h_t is the hypothesis learned at time t and h^* is the best batch hypothesis. In the following, we present several online algorithms for metric learning.

POLA (Shalev-Shwartz et al.) POLA [Shalev-Shwartz et al., 2004], for Pseudo-metric Online Learning Algorithm, is the first online Mahalanobis distance learning approach and learns the matrix $\boldsymbol{M}$ as well as a threshold $b \geq 1$. At each step t, POLA receives a pair $(\boldsymbol{x}_i, \boldsymbol{x}_j, y_{ij})$, where $y_{ij} = 1$ if $(\boldsymbol{x}_i, \boldsymbol{x}_j) \in \mathcal{S}$ and $y_{ij} = -1$ if $(\boldsymbol{x}_i, \boldsymbol{x}_j) \in \mathcal{D}$, and performs two successive orthogonal projections:

1. Projection of the current solution $(\boldsymbol{M}^{t-1}, b^{t-1})$ onto the set $C_1 = \{(\boldsymbol{M}, b) \in \mathbb{R}^{d^2+1} : [y_{ij}(d^2_{\boldsymbol{M}}(\boldsymbol{x}_i, \boldsymbol{x}_j) - b) + 1]_+ = 0\}$, which is done efficiently (closed-form solution). The constraint basically requires that the distance between two instances of same (resp. different) labels be below (resp. above) the threshold b with a margin 1. We get an intermediate solution $(\boldsymbol{M}^{t-\frac{1}{2}}, b^{t-\frac{1}{2}})$ that satisfies this constraint while staying as close as possible to the previous solution.

2. Projection of $(\boldsymbol{M}^{t-\frac{1}{2}}, b^{t-\frac{1}{2}})$ onto the set $C_2 = \{(\boldsymbol{M}, b) \in \mathbb{R}^{d^2+1} : \boldsymbol{M} \in \mathbb{S}^d_+, b \geq 1\}$, which is done rather efficiently (in the worst case, one only needs to compute the minimal eigenvalue of $\boldsymbol{M}^{t-\frac{1}{2}}$). This projects the matrix back onto the PSD cone. We thus get a new solution $(\boldsymbol{M}^t, b^t)$ that yields a valid Mahalanobis distance.

A regret bound for the algorithm is provided. POLA was further improved by Jain et al. [2008], which use LogDet divergence regularization and features tighter regret bounds and more efficient updates. Its practical performance in k-NN classification is similar to (or only slightly worse than) that of ITML.

RDML (Jin et al.) RDML [Jin et al., 2009] is similar to POLA in spirit but is more flexible. At each step t, instead of forcing the margin constraint to be satisfied, it performs a gradient descent step of the following form (assuming Frobenius regularization):

$$\boldsymbol{M}^t = \pi_{\mathbb{S}^d_+}\left(\boldsymbol{M}^{t-1} - \lambda y_{ij}(\boldsymbol{x}_i - \boldsymbol{x}_j)(\boldsymbol{x}_i - \boldsymbol{x}_j)^T\right),$$

where $\pi_{\mathbb{S}^d_+}(\cdot)$ is the projection to the PSD cone. The parameter λ implements a trade-off between satisfying the pairwise constraint and staying close to the previous matrix $\boldsymbol{M}^{t-1}$. Using some linear algebra, the authors show that this update can be performed by solving a convex quadratic program instead of resorting to eigenvalue computation like POLA. RDML is evaluated on several benchmark datasets and is shown to perform comparably to LMNN and ITML in k-NN classification.

OASIS (Chechik et al.) OASIS[8] [Chechik et al., 2009, 2010] learns a bilinear similarity $S_{\boldsymbol{M}}$ instead of a Mahalanobis distance. Since $\boldsymbol{M} \in \mathbb{R}^{d \times d}$ is not required to be PSD, the authors are able to optimize $S_{\boldsymbol{M}}$ in an online manner using a simple and efficient procedure, which belongs to the family of Passive-Aggressive algorithms [Crammer et al., 2006]. The initialization is $\boldsymbol{M} = \boldsymbol{I}$, then at each step t, the algorithm draws a triplet $(\boldsymbol{x}_i, \boldsymbol{x}_j, \boldsymbol{x}_k) \in \mathcal{R}$ and solves the following convex problem:

$$\begin{aligned} \boldsymbol{M}^t = \quad & \underset{\boldsymbol{M},\xi}{\arg\min} \quad \frac{1}{2}\|\boldsymbol{M} - \boldsymbol{M}^{t-1}\|^2_{\mathcal{F}} + C\xi \\ & \text{s.t.} \quad 1 - S_{\boldsymbol{M}}(\boldsymbol{x}_i, \boldsymbol{x}_j) + S_{\boldsymbol{M}}(\boldsymbol{x}_i, \boldsymbol{x}_k) \leq \xi \\ & \qquad \xi \geq 0, \end{aligned} \tag{4.13}$$

[8]Source code available at: `http://ai.stanford.edu/~gal/Research/OASIS/`.

where $C \geq 0$ is the trade-off parameter between minimizing the loss and staying close from the matrix obtained at the previous step. Clearly, if

$$\ell_{M^{t-1}}(\boldsymbol{x}_i, \boldsymbol{x}_j, \boldsymbol{x}_k) = 1 - S_{M^{t-1}}(\boldsymbol{x}_i, \boldsymbol{x}_j) + S_{M^{t-1}}(\boldsymbol{x}_i, \boldsymbol{x}_k) \leq 0,$$

then the solution of (4.13) is $\boldsymbol{M}^t = \boldsymbol{M}^{t-1}$. Otherwise, it is obtained from a simple closed-form update:

$$\begin{aligned} \tau &= \min(C, \ell_{M^{t-1}}(\boldsymbol{x}_i, \boldsymbol{x}_j, \boldsymbol{x}_k)/\|\boldsymbol{V}^t\|^2) \\ \boldsymbol{M}^t &= \boldsymbol{M}^{t-1} + \tau \boldsymbol{V}^t, \end{aligned}$$

where $\boldsymbol{V}^t = \boldsymbol{x}_i(\boldsymbol{x}_j - \boldsymbol{x}_k)^T$ is the gradient matrix.

In practice, OASIS can be applied to problems with millions of training instances and achieves state-of-the-art performance on an image search task using k-NN. However, it cannot use complex regularizers such as sparsity-inducing norms. Note that the same authors derived two other online algorithms as applications of more general frameworks. The first one is based on online learning in the manifold of low-rank matrices [Shalit et al., 2010, 2012] and the second on adaptive regularization of weight matrices [Crammer and Chechik, 2012].

Stochastic and Distributed Optimization

Another popular way to scale to large n is to formulate the problem in a way that can be solved using stochastic and/or distributed optimization. We provide here an informal presentation of how such methods can be used to scale-up metric learning. Interested readers can refer to [Boyd et al., 2011, Dekel et al., 2012, Duchi et al., 2011, 2012, Le Roux et al., 2012, Recht et al., 2011, Shalev-Shwartz and Zhang, 2013] for some recent work on these topics in the context of machine learning.

For the purpose of this section, we assume that our metric learning problem is formulated as the minimization of an average of convex (sub)differentiable functions, one for each training constraint. For instance, in the case of triplet constraints, we would have:

$$\min_{\boldsymbol{M}} \quad \frac{1}{|\mathcal{R}|} \sum_{r=1}^{|\mathcal{R}|} \ell(\boldsymbol{M}, C_r), \tag{4.14}$$

where $C_r = (\boldsymbol{x}_i, \boldsymbol{x}_j, \boldsymbol{x}_k) \in \mathcal{R}$ represents a training triplet.[9] Many of the previously introduced metric learning methods can be reformulated in this way, often by replacing the soft margin constraints by hinge loss terms $[t]_+ = \max(0, t)$ in the objective. For instance, the constrained formulation S&J in (4.6) is equivalent to the unconstrained problem

$$\min_{\boldsymbol{M} \in \mathbb{S}_+^d} \quad \sum_{(\boldsymbol{x}_i, \boldsymbol{x}_j, \boldsymbol{x}_k) \in \mathcal{R}} [1 + d_{\boldsymbol{M}}^2(\boldsymbol{x}_i, \boldsymbol{x}_j) - d_{\boldsymbol{M}}^2(\boldsymbol{x}_i, \boldsymbol{x}_k)]_+ \quad + \quad \lambda \|\boldsymbol{M}\|_{\mathcal{F}}^2.$$

[9]For ease of presentation, we ignore any constraint or regularization on $\boldsymbol{M}$, but these cases can be addressed efficiently with methods such as (stochastic) projected or proximal gradient.

We could use a (sub)gradient descent (GD) algorithm to solve (4.14), however computing the gradient at each step would require $O(|\mathcal{R}|)$ time. One solution is to distribute the training constraints across several machines of a commodity cluster. Each machine then computes its local gradient in parallel and a master node collects and averages them to form the full gradient.

However, the number of training constraints can be so large that it is difficult to efficiently distribute the gradient computation. This could be because there are not enough machines or because communication costs hinder the use of additional machines. In this case, stochastic optimization can be useful. Instead of the full gradient, Stochastic Gradient Descent (SGD) uses an unbiased estimate of the gradient, for instance based on a single term of the sum (4.14) sampled uniformly at random at each iteration. Like full GD, SGD also converges to an optimum (in expectation) under appropriate step size. Its convergence rate is typically slower than that of full GD due to the variance of the gradient estimates, but its very cheap iterations make it very attractive in the large-scale setting [Bottou and Bousquet, 2007].

To reduce the variance, mini-batch SGD picks a random subset of size $m > 1$ instead of a single term to estimate the gradient, providing a middle ground between SGD and full GD. In this context, Clémençon et al. [2015] have shown, both theoretically and empirically, that randomly sampling pairs or triplets directly to build the mini-batch leads to much faster convergence than randomly sampling a subset of points and then forming all possible pairs/triplets based on this subsample. Mini-batch SGD is particularly convenient to combine the benefits of stochastic and distributed optimization, where each machine can compute a local gradient based on a mini-batch. For instance, Xie and Xing [2014] adapt the idea of distributed mini-batch SGD to metric learning, allowing asynchronous gradient updates.

4.3.2 LARGE d: METRIC LEARNING IN HIGH DIMENSIONS

In many application domains such as text and image processing or bioinformatics, data is often represented as very high-dimensional vectors (d is in the order of 10^5 or higher). This is a major problem for classic metric learning algorithms for at least three reasons. First, a single iteration typically requires $O(d^2)$ or even $O(d^3)$ time, which is intractable for large d. Second, merely storing a $d \times d$ matrix may be undesirable, if not infeasible. Lastly, learning $O(d^2)$ parameters is likely to lead to severe overfitting. In this section, we present a few ways to go around the high-dimensionality problem.

Dimensionality Reduction

Virtually all metric learning methods presented earlier (for instance, LMNN and ITML) resort to dimensionality reduction (such as PCA or random projections) to preprocess the data when it has more than a few hundred dimensions, and learn the metric in the reduced space. The resulting metric generally achieves good practical performance (often superior to a standard metric in the original space), but there is a potential loss of performance and interpretability.

Very recently, Qian et al. [2014] proposed a strategy to learn the metric in a reduced space generated by random projections, and yet to be able to recover a full metric in the original space. The approach is promising but the recovery is costly and it is not clear how much the proposed method improves upon simply using the metric learned in the reduced space.

Diagonal Metric

A drastic solution is to learn a diagonal matrix $\boldsymbol{M}$ as done by Schultz and Joachims [2003] (see Section 4.1.2) and more recently by Gao et al. [2014]. This can work well in practice but is very restrictive as it amounts to learning a simple weighting of the features.

Matrix Decompositions

Another strategy is to work with an explicit decomposition of the matrix $\boldsymbol{M}$ to reduce the number of parameters to learn. A popular choice is the low-rank decomposition $\boldsymbol{M} = \boldsymbol{L}\boldsymbol{L}^T$ with $\boldsymbol{L} \in \mathbb{R}^{r \times d}$. However, this leads to nonconvex formulations with possibly many bad local optima, and requires r to be tuned carefully.

Instead, one can decompose $\boldsymbol{M}$ as a sum of rank-one matrices. Unlike BoostMetric (Section 4.1.2), Shi et al. [2014] generate a set of rank-one matrices before training and only need to learn the combining weights. However, a larger dictionary is typically needed as the dimensionality increases, which can be expensive to build and/or cause memory issues. Finally, Liu et al. [2014] build on the observation that when data is very high-dimensional, it is often sparse (i.e., each point has many zero features). In this context, they propose to decompose $\boldsymbol{M}$ as a convex combination of rank-one 4-sparse matrices, each involving only two features. Exploiting the sparsity structure of the problem, they derive a greedy algorithm that incorporates a single such basis into the metric at each iteration, with computational cost independent of d. This allows explicit control of the complexity of the metric to avoid overfitting, and results in extremely sparse high-dimensional metrics that are fast to compute.

4.3.3 LARGE n AND LARGE d

We note that the techniques presented in Section 4.3.1 and Section 4.3.2 may be combined to obtain algorithms that can efficiently deal with simultaneously large n and d.

CHAPTER 5

Nonlinear and Local Metric Learning

As we have seen in the previous chapter, a lot of work has focused on learning a linear metric. Such formulations are convenient to optimize (they are typically convex with the guarantee of finding the global optimum) and the learned metric rarely overfit. For some problems, however, the data has some nonlinear structure that a linear metric is unable to capture. In such cases, a nonlinear metric has the potential to achieve much better performance. In this chapter, we present the two main lines of research in nonlinear metric learning: learn a nonlinear form of metric (Section 5.1) or multiple linear metrics (Section 5.2), as illustrated in Figure 5.1.

5.1 NONLINEAR METHODS

There are two main approaches to nonlinear metric learning. Section 5.1.1 shows how one can learn a linear metric in the feature space induced by a nonlinear kernel. One may also learn a nonlinear metric directly: these approaches are discussed in Section 5.1.2.

5.1.1 KERNELIZATION OF LINEAR METHODS

Inspired by other kernel methods [Schölkopf and Smola, 2001], the motivation for kernelized metric learning is to get the best of both worlds: by learning a linear metric in an implicit nonlinear space, one can combine the expressiveness of a nonlinear metric with the convexity and efficiency of linear metric learning.

The kernel trick for metric learning is derived as follows. Let $K(\boldsymbol{x}, \boldsymbol{x}') = \langle \phi(\boldsymbol{x}), \phi(\boldsymbol{x}') \rangle$ be a kernel, where $\phi(\boldsymbol{x}) \in \mathbb{R}^D$. Let $\{\boldsymbol{x}_i\}_{i=1}^n$ be the training instances and denote $\phi_i = \phi(\boldsymbol{x}_i)$ for simplicity. Consider a (squared) Mahalanobis distance in kernel space:

$$d_{\boldsymbol{M}}^2(\phi_i, \phi_j) = (\phi_i - \phi_j)^T \boldsymbol{M} (\phi_i - \phi_j) = (\phi_i - \phi_j)^T \boldsymbol{L}^T \boldsymbol{L} (\phi_i - \phi_j).$$

Let $\boldsymbol{\Phi} = [\phi_1, \dots, \phi_n]$ and use the parameterization $\boldsymbol{L}^T = \boldsymbol{\Phi} \boldsymbol{U}^T$, where $\boldsymbol{U} \in \mathbb{R}^{D \times n}$. We get:

$$d_{\boldsymbol{M}}^2(\phi(\boldsymbol{x}), \phi(\boldsymbol{x}')) = (\boldsymbol{k} - \boldsymbol{k}')^T \boldsymbol{M} (\boldsymbol{k} - \boldsymbol{k}'), \tag{5.1}$$

where $\boldsymbol{M} \in \mathbb{R}^{n \times n}$, $\boldsymbol{k} = \boldsymbol{\Phi}^T \phi(\boldsymbol{x}) = [K(\boldsymbol{x}_1, \boldsymbol{x}), \dots, K(\boldsymbol{x}_n, \boldsymbol{x})]^T$ and likewise for $\boldsymbol{k}'$. Note that (5.1) can be computed for any two inputs $\boldsymbol{x}, \boldsymbol{x}'$ based on the training instances and the kernel function without ever operating in the D-dimensional kernel space. This allows efficient distance

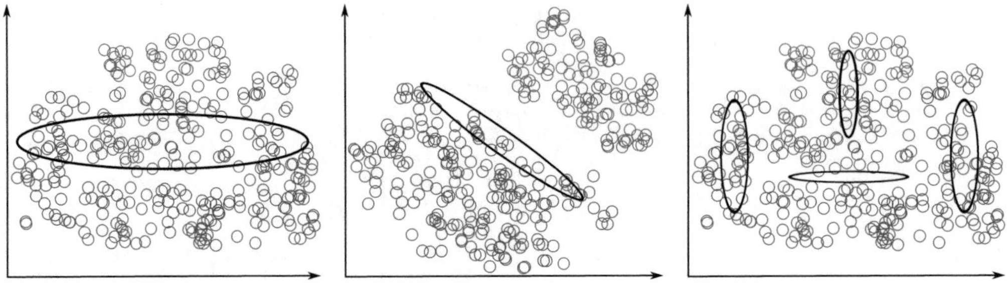

(a) Linear metric (b) Kernelized metric (c) Multiple local metrics

Figure 5.1: When data is not linearly separable, a linear metric cannot capture well the structure of the problem (Figure 5.1(a)). Instead, one can use a kernel to implicitly project data into a nonlinear space where a linear metric performs well (Figure 5.1(b)) or learn multiple linear metrics at several locations of the original feature space (Figure 5.1(c)). Metrics are shown as ellipses representing the set of point at unit distance of the center.

computations in kernel space even when that space is infinite-dimensional (for the Gaussian RBF kernel for instance). Moreover, the distance can be learned by estimating n^2 parameters, which is independent from both the original feature space and the kernel space dimensions. The framework is theoretically supported by a representer theorem [Chatpatanasiri et al., 2010].

In general, kernelizing a metric learning algorithm is however nontrivial, because the interface to the data must be limited to inner products only, so that efficient computation can be achieved through the kernel function. Some metric learning approaches have been shown to be kernelizable using method-specific arguments [see for instance Davis et al., 2007, Hoi et al., 2006, Schultz and Joachims, 2003, Shalev-Shwartz et al., 2004, Torresani and Lee, 2006]. More recently, several authors [Chatpatanasiri et al., 2010, Zhang et al., 2010] have proposed general kernelization results based on Kernel Principal Component Analysis [Schölkopf et al., 1998], a nonlinear extension of PCA [Pearson, 1901]. In short, KPCA implicitly projects the data into the nonlinear (potentially infinite-dimensional) feature space induced by a kernel and performs dimensionality reduction in that space. The (unchanged) linear metric learning algorithm can then be used to learn a metric in the resulting reduced nonlinear space. This is referred to as the "KPCA trick" and can also be used to avoid learning n^2 parameters, which can be difficult when n is large. Chatpatanasiri et al. [2010] showed that the KPCA trick is theoretically sound for unconstrained metric learning algorithms.

Another trick (similar in spirit in the sense that it involves some nonlinear preprocessing of the feature space) is based on kernel density estimation and allows one to deal with both numerical and categorical attributes [He et al., 2013]. General kernelization results can also be obtained from the equivalence between Mahalanobis distance learning in kernel space and linear transformation kernel learning [Jain et al., 2010, 2012], but are restricted to spectral regularizers. Lastly,

Wang et al. [2011] address the problem of choosing an appropriate kernel function by proposing a multiple kernel framework for metric learning.

5.1.2 LEARNING NONLINEAR FORMS OF METRICS

A few approaches have tackled the direct optimization of nonlinear forms of metrics. These approaches are typically nonconvex (thus subject to local optima) but potentially more flexible than kernelized methods, and also more scalable on large datasets.

LSMD (Chopra et al.) Chopra et al. [2005] pioneered the nonlinear metric learning literature. They learn a nonlinear projection $G_{\boldsymbol{W}}(\boldsymbol{x})$ parameterized by a vector $\boldsymbol{W}$ such that the L_2 distance in the low-dimensional target space $d_{\boldsymbol{W}}(\boldsymbol{x}, \boldsymbol{x}') = \|G_{\boldsymbol{W}}(\boldsymbol{x}) - G_{\boldsymbol{W}}(\boldsymbol{x}')\|_2$ is small for positive pairs and large for negative pairs. Precisely, they use the following loss fonction (ignoring constants):

$$\ell(\boldsymbol{W}, \boldsymbol{x}_i, \boldsymbol{x}_j) = (1 - y_{ij}) d^2_{\boldsymbol{W}}(\boldsymbol{x}_i, \boldsymbol{x}_j) + y_{ij} e^{-d_{\boldsymbol{W}}(\boldsymbol{x}_i, \boldsymbol{x}_j)},$$

where $y_{ij} = 0$ if $(\boldsymbol{x}_i, \boldsymbol{x}_j) \in \mathcal{S}$ and 1 if $(\boldsymbol{x}_i, \boldsymbol{x}_j) \in \mathcal{D}$. This choice of loss function is justified by a separation criterion.

No assumption is made about the nature of the nonlinear projection $G_{\boldsymbol{W}}$: the parameter $\boldsymbol{W}$ corresponds to the weights in a deep learning model (more precisely, a multilayer convolutional neural network) and can thus be an arbitrarily complex nonlinear mapping.[1] These weights are learned so as to minimize the loss function through classic neural network techniques, namely back-propagation and stochastic gradient descent. The approach is subject to local optima and requires careful tuning of the many hyperparameters. A significant amount of validation data should thus be used in order to avoid overfitting. Nevertheless, the authors demonstrate the usefulness of LSMD on face verification tasks.

Although deep learning is not yet very well understood and has a number of apparent shortcomings (in particular local optima and a large number of hyperparameters), there has been recent progress in efficiently training such models on very large-scale data, which has led to state-of-the-art performance in many application domains such as computer vision and speech recognition [Deng and Yu, 2014]. Several other metric learning methods based on deep learning have thus been recently proposed. Salakhutdinov and Hinton [2007] use a deep belief network to learn a nonlinear low-dimensional representation of the data and train the last layer to optimize the NCA objective (cf. Section 4.1.1). Hu et al. [2014] propose another deep metric learning architecture to optimize a hinge loss on pairs of instances.

GB-LMNN (Kedem et al.) Kedem et al. [2012] propose Gradient-Boosted LMNN,[2] a generalization of LMNN to learn a distance in a nonlinear projection space defined by ϕ:

$$d_\phi(\boldsymbol{x}, \boldsymbol{x}') = \|\phi(\boldsymbol{x}) - \phi(\boldsymbol{x}')\|_2.$$

[1]An introduction to deep learning is outside the scope of this book. The interested reader may refer to Bengio [2009] and Schmidhuber [2014] for recent surveys.

[2]Source code available at: `http://www.cse.wustl.edu/~kilian/code/code.html`.

Unlike LSMD, this nonlinear mapping takes the form of an additive function

$$\phi(\boldsymbol{x}) = \phi_0(\boldsymbol{x}) + \alpha \sum_{t=1}^{T} h_t(\boldsymbol{x}),$$

where $h_1, \dots, h_T$ are gradient boosted regression trees [Friedman, 2001] of limited depth p and d-dimensional output, $\phi_0(\boldsymbol{x})$ is the linear mapping learned by the original LMNN and α a learning rate.

The authors use the LMNN objective (4.4) reformulated as an unconstrained problem using the hinge loss $[t]_+ = \max(0, t)$. Denoting $\phi_i = \phi(\boldsymbol{x}_i)$:

$$f(\phi) = \sum_{(\boldsymbol{x}_i,\boldsymbol{x}_j) \in \mathcal{S}_{lmnn}} \|\phi_i - \phi_j\|_2^2 + \lambda \sum_{(\boldsymbol{x}_i,\boldsymbol{x}_j,\boldsymbol{x}_k) \in \mathcal{R}_{lmnn}} \left[1 + \|\phi_i - \phi_j\|_2^2 - \|\phi_i - \phi_k\|_2^2\right]_+ .$$

The proposed algorithm aims at iteratively adding a new tree h_t at each step t. Ideally, we would like to select the tree that minimizes the above objective upon its addition to the ensemble:

$$\phi_t(\boldsymbol{x}) = \phi_{t-1}(\boldsymbol{x}) + \alpha h_t(\boldsymbol{x}), \quad \text{where } h_t = \operatorname*{arg\,min}_{h \in \mathcal{T}_p} f(\phi_{t-1} + \alpha h),$$

where $\mathcal{T}_p$ is the set of all regression trees of depth p. To avoid the computational cost of minimizing f at each iteration, h_t is chosen so as to approximate the negative gradient direction of the current objective in the least-square sense:

$$h_t = \operatorname*{arg\,min}_{h \in \mathcal{T}_p} \sum_{i=1}^{n} \|g_t(\boldsymbol{x}_i) + h(\boldsymbol{x}_i)\|_2^2,$$

where $g_t(\boldsymbol{x}_i)$ is the gradient of $f(\phi_{t-1})$ with respect to $\phi_{t-1}(\boldsymbol{x}_i)$. The trees are learned greedily using a classic CART implementation. On an intuitive level, each tree divides the space into 2^p regions, and instances falling in the same region are translated by the same vector—thus examples in different regions are translated in different directions.

Dimensionality reduction can be achieved by learning trees with r-dimensional output. In practice, GB-LMNN appears to be robust to overfitting and often outperforms LMNN and ITML significantly in k-NN classification.

HDML (Norouzi et al.) The originality of Hamming Distance Metric Learning [Norouzi et al., 2012a] is to learn mappings from real-valued vectors to binary codes on which the Hamming distance performs well.[3] Recall that the Hamming distance d_H between two binary codes of same length is simply the number of bits on which they disagree. A great advantage of working with binary codes is their small storage cost and the fact that exact neighbor search can be done in sublinear time [Norouzi et al., 2012b].

[3]Source code available at: `https://github.com/norouzi/hdml`.

The goal here is to optimize a mapping $b(\boldsymbol{x})$ that projects a d-dimensional real-valued input $\boldsymbol{x}$ to a q-dimensional binary code. The mapping takes the general form:

$$b(\boldsymbol{x}; \boldsymbol{w}) = \text{sign}\left(f(\boldsymbol{x}; \boldsymbol{w})\right),$$

where $f : \mathbb{R}^d \to \mathbb{R}^q$ can be any function differentiable in $\boldsymbol{w}$ (e.g., a linear or a nonlinear transformation), sign$(\cdot)$ is the element-wise sign function and $\boldsymbol{w}$ is a real-valued vector representing the parameters to be learned.

Given a relative constraint $(\boldsymbol{x}_i, \boldsymbol{x}_j, \boldsymbol{x}_k) \in \mathcal{R}$, denote by $\boldsymbol{h}_i$, $\boldsymbol{h}_j$ and $\boldsymbol{h}_k$ their corresponding binary codes given by b. The loss is then given by

$$\ell(\boldsymbol{h}_i, \boldsymbol{h}_j, \boldsymbol{h}_k) = \left[1 - d_H\left(\boldsymbol{h}_i, \boldsymbol{h}_k\right) + d_H\left(\boldsymbol{h}_i, \boldsymbol{h}_j\right)\right]_+.$$

In other words, the loss is zero when the Hamming distance between $\boldsymbol{h}_i$ and $\boldsymbol{h}_j$ is at least one bit smaller than the distance between $\boldsymbol{h}_i$ and $\boldsymbol{h}_k$. HDML is formalized as a loss minimization problem with L_2 norm regularization on $\boldsymbol{w}$:

$$\min_{\boldsymbol{w}} \sum_{(\boldsymbol{x}_i, \boldsymbol{x}_j, \boldsymbol{x}_k) \in \mathcal{R}} \ell(b(\boldsymbol{x}_i; \boldsymbol{w}), b(\boldsymbol{x}_j; \boldsymbol{w}), b(\boldsymbol{x}_k; \boldsymbol{w})) + \frac{\lambda}{2}\|\boldsymbol{w}\|_2^2 \quad (5.2)$$

This objective function is nonconvex and discontinuous, and thus optimization is very challenging. To mitigate the discontinuity problem, the authors rely on a continuous upper bound on the loss function which can be used in a stochastic gradient descent algorithm. Specifically, at each iteration, the algorithm draws a random triplet $(\boldsymbol{x}_i, \boldsymbol{x}_j, \boldsymbol{x}_k) \in \mathcal{R}$ and finds the binary codes given by:

$$(\hat{\boldsymbol{h}}_i, \hat{\boldsymbol{h}}_j, \hat{\boldsymbol{h}}_k) = \operatorname*{arg\,max}_{\boldsymbol{h}_i, \boldsymbol{h}_j, \boldsymbol{h}_k} \{\ell(\boldsymbol{h}_i, \boldsymbol{h}_j, \boldsymbol{h}_k) + \boldsymbol{h}_i^T f(\boldsymbol{x}_i) + \boldsymbol{h}_j^T f(\boldsymbol{x}_j) + \boldsymbol{h}_k^T f(\boldsymbol{x}_k)\}.$$

The authors show that this can be done in $O(q^2)$ time, which is efficient as long as the code length q remains small.

Experiments show that relatively short codes obtained by nonlinear mapping are sufficient to achieve few constraint violations, and that a k-NN classifier based on these codes can achieve competitive performance with state-of-the-art classifiers. Neyshabur et al. [2013] later showed that using asymmetric codes can lead to shorter encodings while maintaining similar performance.

5.2 LEARNING MULTIPLE LOCAL METRICS

The methods studied so far learn a single (linear or nonlinear) metric. However, if the data is heterogeneous, the learned metric may not capture well the complexity of the task and it might be beneficial to learn multiple (linear) metrics that vary across the space (e.g., one for each class or for each instance). This can often be seen as approximating the geodesic distance defined by a metric tensor [see Ramanan and Baker, 2011, for a review on this matter]. It is typically crucial that the local metrics be learned simultaneously in order to make them meaningfully comparable and also to alleviate overfitting. Learning multiple metrics has been shown to significantly outperform single-metric methods on some problems, but typically comes at the expense of higher time and memory requirements. Furthermore, they usually do not give rise to a consistent global metric, although some recent work partially addresses this issue [Hauberg et al., 2012, Zhan et al., 2009].

M^2-LMNN (Weinberger & Saul) Multiple Metrics LMNN[4] [Weinberger and Saul, 2008, 2009] is an extension of LMNN to learn several Mahalanobis distances in different parts of the feature space. As a preprocessing step, training data is partitioned in C clusters. These can be obtained either in a supervised way (using class labels) or without supervision (e.g., using K-Means). Let $C(\boldsymbol{x}) \in \{1, \dots, C\}$ denote the cluster associated with an instance $\boldsymbol{x}$. M^2-LMNN learns C metrics $\boldsymbol{M}_1, \dots, \boldsymbol{M}_C$ (one for each cluster) based on LMNN's objective, with the modification that the distance to a target neighbor or an impostor $\boldsymbol{x}$ is measured under its local metric $\boldsymbol{M}_{C(\boldsymbol{x})}$. Specifically, the formulation is as follows:

$$\begin{aligned} \min_{\boldsymbol{M}_1,\dots,\boldsymbol{M}_C \in \mathbb{S}_+^d, \xi \geq 0} \quad & (1-\mu) \sum_{(\boldsymbol{x}_i,\boldsymbol{x}_j) \in \mathcal{S}_{lmnn}} d^2_{\boldsymbol{M}_{C(\boldsymbol{x}_j)}}(\boldsymbol{x}_i, \boldsymbol{x}_j) \quad + \quad \mu \sum_{i,j,k} \xi_{ijk} \\ \text{s.t.} \quad & d^2_{\boldsymbol{M}_{C(\boldsymbol{x}_k)}}(\boldsymbol{x}_i, \boldsymbol{x}_k) - d^2_{\boldsymbol{M}_{C(\boldsymbol{x}_j)}}(\boldsymbol{x}_i, \boldsymbol{x}_j) \geq 1 - \xi_{ijk} \quad \forall (\boldsymbol{x}_i, \boldsymbol{x}_j, \boldsymbol{x}_k) \in \mathcal{R}_{lmnn}. \end{aligned}$$

The problem remains convex and in practice, M^2-LMNN can yield significant improvements over standard LMNN (especially when supervised clustering is used to determine the clusters). However, this comes at the expense of a higher computational cost as there are C times more parameters to learn. It is also prone to overfitting (since each local metric can be overly specific to its region) unless a large validation set is used [Wang et al., 2012c].

PLML (Wang et al.) Wang et al. [2012c] propose PLML,[5] a Parametric Local Metric Learning method where a Mahalanobis metric $d^2_{\boldsymbol{M}_i}$ is learned for each training instance $\boldsymbol{x}_i$:

$$d^2_{\boldsymbol{M}_i}(\boldsymbol{x}_i, \boldsymbol{x}_j) = (\boldsymbol{x}_i - \boldsymbol{x}_j)^T \boldsymbol{M}_i (\boldsymbol{x}_i - \boldsymbol{x}_j).$$

In order to avoid overfitting and excessive training cost, $\boldsymbol{M}_i$ is parameterized to be a weighted linear combination of a few metric bases $\boldsymbol{M}_{b_1}, \dots, \boldsymbol{M}_{b_m}$, where $\boldsymbol{M}_{b_j} \succeq 0$ is associated with an anchor point $\boldsymbol{u}_j$.[6] In other words, $\boldsymbol{M}_i$ is defined as:

$$\boldsymbol{M}_i = \sum_{j=1}^m W_{ib_j} \boldsymbol{M}_{b_j}, \quad W_{ib_j} \geq 0, \quad \sum_{j=1}^m W_{ib_j} = 1,$$

where the nonnegativity of the weights ensures that the combination is PSD. The proposed algorithm first learns these weights based on a manifold assumption, then learns the basis metrics

[4] Source code available at: `http://www.cse.wustl.edu/~kilian/code/code.html`.

[5] Source code available at: `http://cui.unige.ch/~wangjun/papers/PLML.zip`.

[6] In practice, these anchor points are defined as the means of clusters constructed by the K-Means algorithm.

in a second step. More precisely, the weight learning is formulated as follows:

$$\begin{aligned} \min_{\boldsymbol{W}} \quad & \|\boldsymbol{X} - \boldsymbol{W}\boldsymbol{U}\|_{\mathcal{F}}^2 + \lambda_1 \operatorname{tr}(\boldsymbol{W}\boldsymbol{G}) + \lambda_2 \operatorname{tr}(\boldsymbol{W}^T \boldsymbol{L}\boldsymbol{W}) \\ \text{s.t.} \quad & W_{ib_j} \geq 0, \sum_{j=1}^{m} W_{ib_j} = 1 \quad \forall i, b_j, \end{aligned} \tag{5.3}$$

where $\boldsymbol{X} \in \mathbb{R}^{n \times d}$ and $\boldsymbol{U} \in \mathbb{R}^{m \times d}$ are the training data and anchor point matrices respectively, $\boldsymbol{G} \in \mathbb{R}^{m \times n}$ is the Euclidean distance matrix between anchor points and training points, and $\boldsymbol{L} \in \mathbb{R}^{n \times n}$ is the Laplacian matrix constructed using a k-nearest neighbor graph of the training instances. The objective of (5.3) is a trade-off between three terms: (i) each point $\boldsymbol{x}_i$ should be close to its linear approximation $\sum_{j=1}^{m} W_{ib_j} \boldsymbol{u}_j$, (ii) the weighting scheme should be local (i.e., W_{ib_j} should be large if $\boldsymbol{x}_i$ and $\boldsymbol{u}_i$ are similar), and (iii) the weights should vary smoothly over the data manifold (i.e., similar training instances should be assigned similar weights).[7]

Given the weights, the basis metrics $\boldsymbol{M}_{b_1}, \ldots, \boldsymbol{M}_{b_m}$ are then learned based on the following formulation:

$$\begin{aligned} \min_{M_{b_1},\ldots,M_{b_m} \in \mathbb{S}_+^d, \xi \geq 0} \quad & \sum_{i,j,k} \xi_{ijk} + \alpha_1 \sum_{(\boldsymbol{x}_i,\boldsymbol{x}_j) \in \mathcal{S}} \sum_{b_l} W_{ib_l} d^2_{\boldsymbol{M}_{b_l}}(\boldsymbol{x}_i, \boldsymbol{x}_j) + \alpha_2 \sum_{b_l} \|\boldsymbol{M}_{b_l}\|_{\mathcal{F}}^2 \\ \text{s.t.} \quad & \sum_{b_l} W_{ib_l} \left(d^2_{\boldsymbol{M}_{b_l}}(\boldsymbol{x}_i, \boldsymbol{x}_k) - d^2_{\boldsymbol{M}_{b_l}}(\boldsymbol{x}_i, \boldsymbol{x}_j) \right) \geq 1 - \xi_{ijk} \quad \forall (\boldsymbol{x}_i, \boldsymbol{x}_j, \boldsymbol{x}_k) \in \mathcal{R}. \end{aligned}$$

This is essentially an LMNN formulation with additional Frobenius norm regularization. In terms of scalability, the weight learning procedure is fairly efficient, but computing and storing $\boldsymbol{L}$ is very expensive for large datasets. The basis metric learning procedure requires m full eigen-decompositions at each step, making the approach intractable for problems of moderate to high dimensionality. In practice, PLML outperforms M^2-LMNN on the evaluated datasets, and is quite robust to overfitting due to its global manifold regularization. However, it is sensitive to the relevance of the Euclidean distance to assess the similarity between data points and anchor points. Note that PLML has many hyper-parameters, although in the experiments the authors use default values for most of them.

SCML (Shi et al.) SCML (Sparse Compositional Metric Learning), proposed by Shi et al. [2014], is a general framework to learn Mahalanobis metrics as sparse combinations of a set of bases.[8] Like BoostMetric (Section 4.1.2), it relies on a decomposition of $\boldsymbol{M} \in \mathbb{S}_+^d$ as a sum of rank-one matrices:

$$\boldsymbol{M} = \sum_{i=1}^{K} w_i \boldsymbol{b}_i \boldsymbol{b}_i^T, \quad \boldsymbol{w} \geq 0.$$

[7]The weights of a test instance can be learned by optimizing the same trade-off given the weights of the training instances, or simply set to the weights of the nearest training instance.

[8]Source code available at: `http://mloss.org/software/view/553/`.

The basis set $B = \{\boldsymbol{b}_i\boldsymbol{b}_i^T\}_{i=1}^K$ is provided as input to the algorithm and consists of rank-one matrices that are locally discriminative. Such elements can be efficiently generated from the training data at several local regions (the authors propose to cluster the data and apply Fisher discriminant analysis to each cluster).

For learning multiple local metrics, the idea of SCML is to learn a mapping $W : \mathbb{R}^d \to \mathbb{R}_+^K$ that maps any point $\boldsymbol{x} \in \mathbb{R}^d$ to a vector of weights $W(\boldsymbol{x}) \in \mathbb{R}_+^K$ defining its local metric:

$$d_W(\boldsymbol{x}, \boldsymbol{x}') = (\boldsymbol{x} - \boldsymbol{x}')^T \left(\sum_{i=1}^K [W(\boldsymbol{x})]_i \boldsymbol{b}_i \boldsymbol{b}_i^T \right) (\boldsymbol{x} - \boldsymbol{x}'),$$

where $[W(\boldsymbol{x})]_i$ is the i-th entry of $W(\boldsymbol{x})$. This formulation has several advantages. First, learning such a mapping amounts to implicitly learning a local metric not only for each cluster or each training point (like in M^2-LMNN and PLML) but for any $\boldsymbol{x} \in \mathbb{R}^d$. Second, if W is constrained to be smooth, then similar points are assigned similar weights, providing natural "manifold" regularization. The authors propose the following parameterization:

$$[W(\boldsymbol{x})]_i = \left(\boldsymbol{a}_i^T P(\boldsymbol{x}) + c_i\right)^2,$$

where $\boldsymbol{A} = [\boldsymbol{a}_1 \ldots \boldsymbol{a}_K]^T$ is a $d' \times K$ real-valued matrix, $\boldsymbol{c} \in \mathbb{R}^K$ and $P : \mathbb{R}^d \to \mathbb{R}^{d'}$ is a smooth mapping. Taking the square guarantees that the weights are nonnegative $\forall \boldsymbol{x} \in \mathbb{R}^d$. Note that the number of parameters to learn is $K(d' + 1)$. In principle, P could be the identity mapping, but the authors use a kernel PCA projection so that the weights can vary nonlinearly and to allow dimensionality reduction ($d' < d$) to reduce the number of parameters to learn. The learning problem is formulated as follows:

$$\min_{\tilde{\boldsymbol{A}} \in \mathbb{R}^{(d'+1)\times K}} \frac{1}{|\mathcal{R}|} \sum_{(\boldsymbol{x}_i, \boldsymbol{x}_j, \boldsymbol{x}_k) \in \mathcal{R}} \left[1 + d_W(\boldsymbol{x}_i, \boldsymbol{x}_j) - d_W(\boldsymbol{x}_i, \boldsymbol{x}_k)\right]_+ \quad + \quad \lambda \|\tilde{\boldsymbol{A}}\|_{2,1}, \tag{5.4}$$

where $[t]_+ = \max(0, t)$ is the hinge loss, $\|\cdot\|_{2,1}$ is the mixed $L_{2,1}$ norm used in SML (see Section 4.1.2) and $\tilde{\boldsymbol{A}}$ denotes the concatenation of $\boldsymbol{A}$ and $\boldsymbol{c}$. The $L_{2,1}$ norm introduces sparsity at the column level, regularizing the local metrics to use the same basis subset. Problem (5.4) is nonconvex and is thus subject to local minima. The authors ensure a good solution is found by initializing the parameters to the optimal solution of their single-metric formulation (which is convex). The problem is solved using stochastic optimization (see Section 4.3.1) and can thus deal with a large number of triplet constraints. In practice, SCML is shown to be superior to M^2-LMNN and PLML in k-NN classification on several benchmark datasets.

Bk-means (Wu et al.) Wu et al. [2009, 2012] propose to learn Bregman distances (or Bregman divergences), a family of metrics that do not necessarily satisfy the triangle inequality or symmetry [Bregman, 1967]. Given the strictly convex and twice differentiable function $\varphi : \mathbb{R}^d \to \mathbb{R}$, the Bregman distance is defined as:

$$d_\varphi(\boldsymbol{x}, \boldsymbol{x}') = \varphi(\boldsymbol{x}) - \varphi(\boldsymbol{x}') - (\boldsymbol{x} - \boldsymbol{x}')^T \nabla\varphi(\boldsymbol{x}').$$

It generalizes many widely used measures: the Mahalanobis distance is recovered by setting $\varphi(\boldsymbol{x}) = \frac{1}{2}\boldsymbol{x}^T \boldsymbol{M}\boldsymbol{x}$, the KL divergence [Kullback and Leibler, 1951] by choosing $\varphi(\boldsymbol{p}) = \sum_{i=1}^{d} p_i \log p_i$ (here, $\boldsymbol{p}$ is a discrete probability distribution), etc. The authors consider the following symmetrized version:

$$\begin{aligned} d_\varphi(\boldsymbol{x}, \boldsymbol{x}') &= \left(\nabla\varphi(\boldsymbol{x}) - \nabla\varphi(\boldsymbol{x}')\right)^T (\boldsymbol{x} - \boldsymbol{x}') \\ &= (\boldsymbol{x} - \boldsymbol{x}')^T \nabla^2 \varphi(\tilde{\boldsymbol{x}})(\boldsymbol{x} - \boldsymbol{x}'), \end{aligned}$$

where $\tilde{\boldsymbol{x}}$ is a point on the line segment between $\boldsymbol{x}$ and $\boldsymbol{x}'$. Therefore, d_φ amounts to a Mahalanobis distance parameterized by the Hessian matrix of φ which depends on the location of $\boldsymbol{x}$ and $\boldsymbol{x}'$. In this respect, learning φ can be seen as learning an infinite number of local Mahalanobis distances. They take a nonparametric approach by assuming ϕ to belong to a Reproducing Kernel Hilbert Space $\mathcal{H}_K$ associated to a kernel function $K(\boldsymbol{x}, \boldsymbol{x}') = h(\boldsymbol{x}^T \boldsymbol{x}')$ where $h(z)$ is a strictly convex function (set to $\exp(z)$ in the experiments). This allows the derivation of a representer theorem. Setting $\varphi(\boldsymbol{x}) = \sum_{i=1}^{n} \alpha_i h(\boldsymbol{x}_i^T \boldsymbol{x})$ leads to the following formulation based on classic positive/negative pairs:

$$\min_{\boldsymbol{\alpha}\in\mathbb{R}_+^n, b} \quad \frac{1}{2}\boldsymbol{\alpha}^T \boldsymbol{K}\boldsymbol{\alpha} \quad + \quad C \sum_{(\boldsymbol{x}_i, \boldsymbol{x}_j)\in\mathcal{S}\cup\mathcal{D}} \left[1 - y_{ij}\left(d_\varphi(\boldsymbol{x}_i, \boldsymbol{x}_j) - b\right)\right]_+, \tag{5.5}$$

where $\boldsymbol{K}$ is the Gram matrix, $[t]_+ = \max(0, t)$ is the hinge loss and C is the trade-off parameter. Problem (5.5) is solved by a simple subgradient descent approach where each iteration has a computational complexity linear in the number of training pairs. Note that (5.5) only has $n+1$ variables instead of d^2 in most metric learning formulations, which can be beneficial when $n < d^2$. On the other hand, the approach does not scale well to large datasets: it requires the computation of the $n \times n$ matrix $\boldsymbol{K}$, and computing the learned distance requires n kernel evaluations. The method is evaluated on clustering problems and exhibits good performance, matching or improving that of other metric learning approaches such as ITML or BoostMetric.

RFD (Xiong et al.) The originality of the Random Forest Distance [Xiong et al., 2012] is to cast the metric learning problem as a pair classification problem.[9] Each pair of examples $(\boldsymbol{x}, \boldsymbol{x}')$ is mapped to the following feature space:

$$\phi(\boldsymbol{x}, \boldsymbol{x}') = \begin{bmatrix} |\boldsymbol{x} - \boldsymbol{x}'| \\ \frac{1}{2}(\boldsymbol{x} + \boldsymbol{x}') \end{bmatrix} \in \mathbb{R}^{2d}.$$

The first part of $\phi(\boldsymbol{x}, \boldsymbol{x}')$ encodes the relative position of the examples and the second part their absolute position, as opposed to the implicit mapping of the Mahalanobis distance which only encodes relative information. The learned metric is based on a random forest F [Breiman, 2001],

[9] Source code available at: `http://www.cse.buffalo.edu/~cxiong/RFD_Package.zip`.

i.e., an ensemble of decision trees:

$$d_{RFD}(\boldsymbol{x}, \boldsymbol{x}') = F(\phi(\boldsymbol{x}, \boldsymbol{x}')) = \frac{1}{T}\sum_{t=1}^{T} f_t(\phi(\boldsymbol{x}, \boldsymbol{x}')),$$

where $f_t(\cdot) \in \{0, 1\}$ is the output of decision tree t. RFD is thus highly nonlinear and is able to implicitly adapt the metric throughout the space: when a decision tree in F selects a node split based on a value of the absolute position part, then the entire sub-tree is specific to that region of $\mathbb{R}^{2d}$. As compared to other local metric learning methods, training is very efficient: each tree takes $O(n \log n)$ time to generate and trees can be constructed in parallel. A drawback is that a distance evaluation requires computing the output of the T trees. The experiments highlight the importance of encoding absolute information, and show that RFD is fast to train and clearly outperforms some single metric learning methods (LMNN, ITML) on several datasets in k-NN classification.

CHAPTER 6

Metric Learning for Special Settings

In the previous two chapters, we have reviewed methods to learn metrics in a standard supervised classification setup. In this chapter, we present extensions of metric learning to more complex settings that require specific formulations. Section 6.1 deals with multi-task and transfer learning, while Section 6.2 is devoted to learning a metric for ranking. Section 6.3 addresses the problem of learning a metric from semi-supervised data or in a domain adaptation setting. Finally, Section 6.4 presents approaches to learn metrics specifically for histogram data.

6.1 MULTI-TASK AND TRANSFER LEARNING

In the multi-task learning framework [Caruana, 1997], we consider a set of T different but related tasks, each of these modeled by a probability distribution over $\mathcal{X} \times \mathcal{Y}$. The learner has access to T learning samples, one for each task, and the objective is to improve the performance on all the tasks at the same time by making use of the existing relationships between the tasks. Multi-task transfer learning corresponds to the case where the goal is to improve only one target task with the help of the others. In this section, we cover several Mahalanobis distance learning methods for the multi-task setting where the general trend consists in learning a metric for each task in a coupled fashion in order to improve the performance on all tasks.

mt-LMNN (Parameswaran & Weinberger) Multi-Task LMNN[1] [Parameswaran and Weinberger, 2010] is a straightforward adaptation of the ideas of Multi-Task SVM [Evgeniou and Pontil, 2004] to metric learning. Given T related tasks, they model the problem as learning a shared Mahalanobis metric $d_{\boldsymbol{M}_0}$ as well as task-specific metrics $d_{\boldsymbol{M}_1}, \dots, d_{\boldsymbol{M}_t}$ and define the metric for task t as

$$d_t(\boldsymbol{x}, \boldsymbol{x}') = (\boldsymbol{x} - \boldsymbol{x}')^T (\boldsymbol{M}_0 + \boldsymbol{M}_t)(\boldsymbol{x} - \boldsymbol{x}').$$

Note that $\boldsymbol{M}_0 + \boldsymbol{M}_t \succeq 0$, hence d_t is a valid pseudo-metric. The LMNN formulation (see Section 4.1.1) is easily generalized to this multi-task setting so as to learn the metrics jointly, with a specific regularization term defined as follows:

$$\gamma_0 \|\boldsymbol{M}_0 - \boldsymbol{I}\|_{\mathcal{F}}^2 + \sum_{t=1}^{T} \gamma_t \|\boldsymbol{M}_t\|_{\mathcal{F}}^2,$$

[1]Source code available at: `http://www.cse.wustl.edu/~kilian/code/code.html`.

where γ_t controls the regularization of $\boldsymbol{M}_t$. When $\gamma_0 \to \infty$, the shared metric $d_{\boldsymbol{M}_0}$ is simply the Euclidean distance, and the formulation reduces to T independent LMNN formulations. On the other hand, when $\gamma_{t>0} \to \infty$, the task-specific matrices are simply zero matrices and the formulation reduces to LMNN on the union of all data. In-between these extreme cases, these parameters can be used to adjust the relative importance of each metric: γ_0 to set the overall level of shared information, and γ_t to set the importance of $\boldsymbol{M}_t$ with respect to the shared metric. The formulation remains convex and can be solved using the same solver as LMNN (but there is T times more parameters to learn). In the multi-task setting with limited training data for each task, mt-LMNN clearly outperforms single-task LMNN and other multi-task classification techniques such as mt-SVM.

MLCS (Yang et al.) MLCS (Metric Learning based on Common Subspace) [Yang et al., 2011] is a different approach to the problem of multi-task metric learning. For each task $t \in \{1, \ldots, T\}$, the authors consider learning a Mahalanobis metric

$$d^2_{\boldsymbol{L}_t^T \boldsymbol{L}_t}(\boldsymbol{x}, \boldsymbol{x}') = (\boldsymbol{x} - \boldsymbol{x}')^T \boldsymbol{L}_t^T \boldsymbol{L}_t (\boldsymbol{x} - \boldsymbol{x}') = (\boldsymbol{L}_t \boldsymbol{x} - \boldsymbol{L}_t \boldsymbol{x}')^T (\boldsymbol{L}_t \boldsymbol{x} - \boldsymbol{L}_t \boldsymbol{x}')$$

parameterized by the transformation matrix $\boldsymbol{L}_t \in \mathbb{R}^{r \times d}$. They show that $\boldsymbol{L}_t$ can be decomposed into a "subspace" part $\boldsymbol{L}_0^t \in \mathbb{R}^{r \times d}$ and a "low-dimensional metric" part $\boldsymbol{R}_t \in \mathbb{R}^{r \times r}$ such that $\boldsymbol{L}_t = \boldsymbol{R}_t \boldsymbol{L}_0^t$. The main assumption of MLCS is that all tasks share a common subspace, i.e., $\forall t$, $\boldsymbol{L}_0^t = \boldsymbol{L}_0$. This parameterization can be used to extend most existing metric learning methods to the multi-task setting, although it breaks the convexity of the formulation and is thus subject to local optima. However, as opposed to mt-LMNN, it can be made low-rank by setting $r < d$ and thus has many less parameters to learn. In their work, MLCS is applied to an alternative algorithm to solve LMNN proposed by Torresani and Lee [2006]. The resulting method is evaluated on problems with very scarce training data and study the performance for different values of r. It is shown to outperform mt-LMNN, but the setup is a bit unfair since mt-LMNN is forced to be low-rank by eigenvalue thresholding.

GPML (Yang et al.) The work of Yang et al. [2012] identifies two drawbacks of previous multi-task metric learning approaches: (i) MLCS's assumption of common subspace is sometimes too strict and leads to a nonconvex formulation, and (ii) the Frobenius regularization of mt-LMNN does not preserve geometry. This property is defined as being the ability to propagate side-information: the task-specific metrics should be regularized so as to preserve the relative distance between training pairs. They introduce GPML (Geometry Preserving Metric Learning), which extends any metric learning algorithm to the multi-task setting:

$$\min_{\boldsymbol{M}_0, \ldots, \boldsymbol{M}_t \in \mathbb{S}_+^d} \quad \sum_{i=1}^{t} \left(\ell(\boldsymbol{M}_t, \mathcal{S}_t, \mathcal{D}_t, \mathcal{R}_t) + \gamma d_\varphi(\boldsymbol{M}_t, \boldsymbol{M}_0) \right) \quad + \quad \gamma_0 d_\varphi(\boldsymbol{A}_0, \boldsymbol{M}_0), \tag{6.1}$$

where $\ell(\boldsymbol{M}_t, \mathcal{S}_t, \mathcal{D}_t, \mathcal{R}_t)$ is the loss function for the task t based on the training pairs/triplets (depending on the chosen algorithm), $d_\varphi(\boldsymbol{A}, \boldsymbol{B}) = \varphi(\boldsymbol{A}) - \varphi(\boldsymbol{B}) - \operatorname{tr}\left((\nabla\varphi \boldsymbol{B})^T (\boldsymbol{A} - \boldsymbol{B})\right)$ is a

Bregman matrix divergence [Dhillon and Tropp, 2007] and $\boldsymbol{A}_0$ is a predefined metric (e.g., the identity matrix $\boldsymbol{I}$). mt-LMNN can essentially be recovered from (6.1) by setting $\varphi(\boldsymbol{A}) = \|\boldsymbol{A}\|_{\mathcal{F}}^2$ and additional constraints $\boldsymbol{M}_t \succeq \boldsymbol{M}_0$. The authors focus on the von Neumann divergence:

$$d_{VN}(\boldsymbol{A}, \boldsymbol{B}) = \operatorname{tr}(\boldsymbol{A} \log \boldsymbol{A} - \boldsymbol{A} \log \boldsymbol{B} - \boldsymbol{A} + \boldsymbol{B}),$$

where $\log \boldsymbol{A}$ is the matrix logarithm of $\boldsymbol{A}$. Like the LogDet divergence mentioned earlier (Section 4.1.2), the von Neumann divergence is known to be rank-preserving and to provide automatic enforcement of positive-semidefiniteness. The authors further show that minimizing this divergence encourages geometry preservation between the learned metrics. Problem (6.1) remains convex as long as the original algorithm used for solving each task is convex, and can be solved efficiently using gradient descent methods. In the experiments, the method is adapted to LMNN and outperforms single-task LMNN as well as mt-LMNN, especially when training data is very scarce.

TML (Zhang & Yeung) Zhang and Yeung [2010] propose a transfer metric learning (TML) approach.[2] They assume that we are given S independent source tasks with enough labeled data and that a Mahalanobis distance $\boldsymbol{M}_s$ has been learned for each task s. The goal is to leverage the information of the source metrics to learn a distance $\boldsymbol{M}_t$ for a target task, for which we only have a small amount n_t of labeled data. No assumption is made about the relation between the source tasks and the target task: they may be positively/negatively correlated or uncorrelated. The problem is formulated as follows:

$$\begin{aligned} \min_{\boldsymbol{M}_t \in \mathbb{S}_+^d, \boldsymbol{\Omega} \succeq 0} \quad & \frac{2}{n_t^2} \sum_{i<j} \ell\left(1 - y_i y_j \left[1 - d_{\boldsymbol{M}_t}^2(\boldsymbol{x}_i, \boldsymbol{x}_j)\right]\right) + \frac{\lambda_1}{2} \|\boldsymbol{M}_t\|_{\mathcal{F}}^2 + \frac{\lambda_2}{2} \operatorname{tr}(\tilde{\boldsymbol{M}} \boldsymbol{\Omega}^{-1} \tilde{\boldsymbol{M}}^T) \\ \text{s.t.} \quad & \operatorname{tr}(\boldsymbol{\Omega}) = 1, \end{aligned} \tag{6.2}$$

where $\ell(t) = \max(0, t)$ is the hinge loss, $\tilde{\boldsymbol{M}} = (\operatorname{vec}(\boldsymbol{M}_1), \dots, \operatorname{vec}(\boldsymbol{M}_s), \operatorname{vec}(\boldsymbol{M}_t))$. The first two terms are standard (a loss on all possible pairs and Frobenius regularization) while the third one models the relation between tasks based on a positive definite covariance matrix $\boldsymbol{\Omega}$. Assuming that the source tasks are independent and of equal importance, $\boldsymbol{\Omega}$ can be expressed as

$$\boldsymbol{\Omega} = \begin{pmatrix} \alpha \boldsymbol{I}^{(m-1)\times(m-1)} & \boldsymbol{\omega}_m \\ \boldsymbol{\omega}_m & \omega \end{pmatrix},$$

where $\boldsymbol{\omega}_m$ denotes the task covariances between the target task and the source tasks, and ω denotes the variance of the target task. Problem (6.2) is convex and is solved using an alternating procedure that is guaranteed to converge to the global optimum: (i) fixing $\boldsymbol{\Omega}$ and solving for $\boldsymbol{M}_t$, which is done online with an algorithm similar to RDML, and (ii) fixing $\boldsymbol{M}_t$ and solving for $\boldsymbol{\Omega}$, leading to a second-order cone program whose number of variables and constraints is linear in the number of tasks. In practice, TML consistently outperforms metric learning methods without transfer when training data is scarce.

[2]Source code available at: `http://www.cse.ust.hk/~dyyeung/`.

6.2 LEARNING TO RANK

Learning to rank is a recent topic in machine learning with applications in information retrieval, collaborative filtering and ad placement among others [Liu, 2009]. The goal is to learn how to rank items (for instance by relevance to a query) based on some (partial) order on the training data. There are various performance criteria in ranking. For example, the Area Under the ROC Curve (AUC) corresponds to the probability that a classifier will rank a randomly chosen positive instance higher than a randomly chosen negative one, while the Precision-at-k is only concerned with the quality of the ranking of the top k items.

MLR (McFee & Lankriet) The idea of MLR (Metric Learning to Rank) [McFee and Lanckriet, 2010] is to learn a metric for ranking items with respect to queries.[3] Using a distance function between the query and an item as the score to rank the items is a natural approach. Let $\mathcal{P}$ be the set of all permutations (i.e., possible rankings) over the training set. Given a Mahalanobis distance $d_{\boldsymbol{M}}^2$ and a query $\boldsymbol{q}$ (which can be seen as an element of the instance space), the predicted ranking $p \in \mathcal{P}$ consists in sorting the instances by ascending $d_{\boldsymbol{M}}^2(\boldsymbol{q}, \cdot)$. The idea is that if an instance $\boldsymbol{x}_i$ is more relevant than another $\boldsymbol{x}_j$ with respect to $\boldsymbol{q}$, then one must have $d_{\boldsymbol{M}}^2(\boldsymbol{q}, \boldsymbol{x}_i) < d_{\boldsymbol{M}}^2(\boldsymbol{q}, \boldsymbol{x}_j)$. The approach to learn $\boldsymbol{M}$ is based on Structural SVM [Tsochantaridis et al., 2005]. It works directly on $d_{\boldsymbol{M}}^2$ but rather use a feature map encoding the relation between a query and instance rankings as follows:

$$\begin{aligned} \min_{\boldsymbol{M} \in \mathbb{S}_+^d} \quad & \|\boldsymbol{M}\|_* \quad + \quad C \sum_i \xi_i \\ \text{s.t.} \quad & \langle \boldsymbol{M}, \psi(\boldsymbol{x}_i, p_i) - \psi(\boldsymbol{x}_i, p) \rangle_{\mathcal{F}} \geq \Delta(p_i, p) - \xi_i \quad \forall i \in \{1, \ldots, n\}, p \in \mathcal{P}, \end{aligned} \tag{6.3}$$

where $\|\boldsymbol{M}\|_* = \text{tr}(\boldsymbol{M})$ is the nuclear norm, $C \geq 0$ the trade-off parameter, $\langle \boldsymbol{A}, \boldsymbol{B} \rangle_{\mathcal{F}} = \sum_{i,j} A_{ij} B_{ij}$ the Frobenius inner product, $\psi : \mathbb{R}^d \times \mathcal{P} \to \mathbb{S}^d$ the feature encoding of an input-output pair $(\boldsymbol{x}_i, p)$,[4] and $\Delta(p_i, p) \in [0, 1]$ the "margin" representing the loss of predicting ranking p instead of the true ranking p_i. In other words, $\Delta(p_i, p)$ assesses the quality of ranking p with respect to the best ranking p_i and can be evaluated using several measures, such as the AUC, Precision-at-k or Mean Average Precision.

Since the number of constraints is super-exponential in the number of training instances, the authors solve (6.3) using a 1-slack cutting-plane approach [Joachims et al., 2009] which essentially iteratively optimizes over a small set of active constraints (adding the most violated ones at each step) using subgradient descent. However, the algorithm requires a full eigendecomposition of $\boldsymbol{M}$ at each iteration, incurring an $O(d^3)$ cost. In practice, it is competitive with other metric learning algorithms for k-NN classification and a structural SVM algorithm for ranking, and can induce low-rank solutions thanks to the nuclear norm regularization.

[3] Source code is available at: `http://www-cse.ucsd.edu/~bmcfee/code/mlr`.

[4] The feature map ψ is designed such that the ranking p which maximizes $\langle \boldsymbol{M}, \psi(\boldsymbol{q}, p) \rangle_{\mathcal{F}}$ is the one given by ascending $d_{\boldsymbol{M}}^2(\boldsymbol{q}, \cdot)$.

In follow-up work, Lim and Lanckriet [2014] use simpler Frobenius norm regularization and provide a more efficient algorithm. Lim et al. [2013] propose R-MLR, an extension to MLR to deal with the presence of noisy features using the mixed $L_{2,1}$ norm as in SML (see Section 4.1.2). R-MLR is shown to be able to ignore most of the irrelevant features and outperforms MLR in this situation.

6.3 SEMI-SUPERVISED LEARNING

In this section, we present two categories of metric learning methods that are designed to deal with semi-supervised learning tasks. The first one corresponds to the standard semi-supervised setting, where the learner makes use of unlabeled pairs in addition to positive and negative constraints. The second one deals with semi-supervised domain adaptation problems where the learner has access to labeled data drawn according to a source distribution and unlabeled data generated from a different (but related) target distribution.

6.3.1 CLASSIC SETTING

The metric learning methods presented in this section leverage the information brought by the set of *unlabeled pairs*, i.e., pairs of training examples that do not belong to the sets of positive and negative pairs:

$$\mathcal{U} = \{(\boldsymbol{x}_i, \boldsymbol{x}_j) : i \neq j, (\boldsymbol{x}_i, \boldsymbol{x}_j) \notin \mathcal{S} \cup \mathcal{D}\}.$$

An early approach by Bilenko et al. [2004] combined semi-supervised clustering with metric learning. In the following, we review general metric learning formulations that incorporate information from the set of unlabeled pairs $\mathcal{U}$.

LRML (Hoi et al.) Hoi et al. [2008, 2010] propose to follow the principles of manifold regularization for semi-supervised learning [Belkin and Niyogi, 2004] by resorting to a weight matrix $\boldsymbol{W}$ that encodes the similarity between pairs of points.[5] Hoi et al. construct $\boldsymbol{W}$ using the Euclidean distance as follows:

$$W_{ij} = \begin{cases} 1 & \text{if } \boldsymbol{x}_i \in \mathcal{N}(\boldsymbol{x}_j) \text{ or } \boldsymbol{x}_j \in \mathcal{N}(\boldsymbol{x}_i) \\ 0 & \text{otherwise} \end{cases}$$

where $\mathcal{N}(\boldsymbol{x}_j)$ denotes the nearest neighbor list of $\boldsymbol{x}_j$. Using $\boldsymbol{W}$, they use the following regularization known as the graph Laplacian regularizer:

$$\frac{1}{2}\sum_{i,j=1}^{n} d_M^2(\boldsymbol{x}_i, \boldsymbol{x}_j) W_{ij} = \text{tr}(\boldsymbol{X}\boldsymbol{L}\boldsymbol{X}^T\boldsymbol{M}), \tag{6.4}$$

where $\boldsymbol{X}$ is the data matrix and $\boldsymbol{L} = \boldsymbol{D} - \boldsymbol{W}$ is the graph Laplacian matrix with $\boldsymbol{D}$ a diagonal matrix such that $D_{ii} = \sum_j W_{ij}$. Intuitively, (6.4) favors an "affinity-preserving" metric: the distance between points that are similar according to $\boldsymbol{W}$ should remain small according to the learned

[5] Source code available at: `http://www.ee.columbia.edu/~wliu/`.

metric. This regularizer can be added to any metric learning algorithm. Experiments show that LRML (Laplacian Regularized Metric Learning) significantly outperforms supervised methods when the side information is scarce. An obvious drawback is that computing $\boldsymbol{W}$ exactly may be intractable for large-scale datasets. This work has inspired a number of extensions and improvements: Liu et al. [2010] introduce a refined way of constructing $\boldsymbol{W}$ while Baghshah and Shouraki [2009], Zhong et al. [2011] and Wang et al. [2013a] use a different (but similar in spirit) manifold regularizer.

M-DML (Zha et al.) The idea of Zha et al. [2009] is to augment the Laplacian regularization (6.4) using metrics $\boldsymbol{M}_1, \ldots, \boldsymbol{M}_K$ learned from auxiliary datasets. Formally, for each available auxiliary metric, a weight matrix $\boldsymbol{W}_k$ is constructed following Hoi et al. [2008, 2010] but using metric $\boldsymbol{M}_k$ instead of the Euclidean distance. These are then combined to obtain the following regularizer:

$$\sum_{k=1}^{K} \alpha_k \operatorname{tr}(\boldsymbol{X}\boldsymbol{L}_k\boldsymbol{X}^T\boldsymbol{M}),$$

where $\boldsymbol{L}_k$ is the Laplacian matrix associated with weight matrix $\boldsymbol{W}_k$ and α_k is the weight reflecting the utility of auxiliary metric $\boldsymbol{M}_k$. As such weights are difficult to set in practice, Zha et al. propose to learn them together with the metric $\boldsymbol{M}$ by alternating optimization (which only converges to a local minimum). Experiments on a face recognition task show that metrics learned from auxiliary datasets can be successfully used to improve performance over LRML.

6.3.2 DOMAIN ADAPTATION

In the context of supervised learning, domain adaptation (DA) [Ben-David et al., 2010, Mansour et al., 2009, Quiñonero-Candela, 2009] refers to a setting where the labeled training data and the test data come from different (but somehow related) distributions over $\mathcal{X} \times \mathcal{Y}$, referred to as the source p_S and target p_T distributions respectively. This situation occurs very often in real-world applications (such as speech recognition, spam detection and object recognition). The learner is generally provided a sample of labeled examples coming from the source distribution with a sample of unlabeled examples and/or few labeled instances coming from the target distribution. The objective is then to learn a good predictive model on the target domain with the help of the data coming from the source domain. This is generally tackled by trying to move closer the two distributions while ensuring a good predictive performance on the source domain. Metric learning can provide interesting solutions to this problem. The first method presented in this section requires a small sample of target labeled data, while the others address the more challenging case where only a sample unlabeled target data is available.

ARC-t (Kulis et al.) ARC-t [Kulis et al., 2011], for Asymmetric Regularized Cross-domain transformation, is a domain adaptation approach requiring a labeled source sample and a (small) labeled target sample. It optimizes a bilinear similarity $S_{\boldsymbol{M}}(\boldsymbol{x}, \boldsymbol{y}) = \boldsymbol{x}^T\boldsymbol{M}\boldsymbol{y}$ where $\boldsymbol{x} \in \mathbb{R}^s$ belongs to the source domain, $\boldsymbol{y} \in \mathbb{R}^t$ belongs to the target domain and $\boldsymbol{M} \in \mathbb{R}^{s \times t}$. Note that $\boldsymbol{x}$ and $\boldsymbol{y}$ do not need to have the same dimensionality, in which case $\boldsymbol{M}$ is a rectangular matrix.

The sets of similar (resp. dissimilar) pairs are constructed by pairing source and target examples that have the same (resp. different) label. The goal is essentially to map points from one domain to another. The problem formulation is quite straightforward:

$$\min_{\boldsymbol{M} \in \mathbb{R}^{s \times t}} \sum_{(\boldsymbol{x},\boldsymbol{y}) \in \mathcal{S}} \max(0, u - S_{\boldsymbol{M}}(\boldsymbol{x}, \boldsymbol{y})) + \sum_{(\boldsymbol{x},\boldsymbol{y}) \in \mathcal{D}} \max(0, S_{\boldsymbol{M}}(\boldsymbol{x}, \boldsymbol{y}) - v) + \lambda \|\boldsymbol{M}\|_{\mathcal{F}}. \tag{6.5}$$

A kernelized version is also proposed. The proposed method is shown to outperform several domain adaptation baselines on object recognition tasks.

This method is an extension of a previous approach [Saenko et al., 2010] which optimizes a Mahalanobis distance instead of a similarity based on ITML formulation. This formulation is restricted to the case where $\boldsymbol{x}$ and $\boldsymbol{y}$ have the same dimensionality.

CDML (Cao et al.) CDML [Cao et al., 2011], for Consistent Distance Metric Learning, deals with the setting of covariate shift, which assumes that source and target data distributions $p_S(\boldsymbol{x})$ and $p_T(\boldsymbol{x})$ are different but the conditional distribution of the labels given the features remains the same, i.e., $p_S(y|\boldsymbol{x}) = p_T(y|\boldsymbol{x})$. In the context of metric learning, the assumption is made at the pair level, i.e., $p(y_{ij}|\boldsymbol{x}_i, \boldsymbol{x}_j)$ is stable across domains S and T. Cao et al. show that if some metric learning algorithm minimizing some training loss $\sum_{(\boldsymbol{x}_i,\boldsymbol{x}_j) \in \mathcal{S} \cup \mathcal{D}} \ell(d^2_{\boldsymbol{M}}, \boldsymbol{x}_i, \boldsymbol{x}_j, y_{ij})$ is asymptotically consistent without covariate shift, then the following algorithm is consistent under covariate shift:

$$\min_{\boldsymbol{M} \in \mathbb{S}^d_+} \sum_{(\boldsymbol{x}_i,\boldsymbol{x}_j) \in \mathcal{S} \cup \mathcal{D}} w_{ij} \ell(d^2_{\boldsymbol{M}}, \boldsymbol{x}_i, \boldsymbol{x}_j, y_{ij}), \quad \text{where } w_{ij} = \frac{p_T(\boldsymbol{x}_i) p_T(\boldsymbol{x}_j)}{p_S(\boldsymbol{x}_i) p_S(\boldsymbol{x}_j)}. \tag{6.6}$$

Problem (6.6) can be seen as cost-sensitive metric learning, where the cost of each pair is given by the importance weight w_{ij}. Therefore, adapting a metric learning algorithm to covariate shift boils down to computing the importance weights, which can be done reliably using unlabeled data from the target domain [Tsuboi et al., 2008]. The authors experiment with ITML and show that their adapted version outperforms standard ITML in situations of (real or simulated) covariate shift.

DAML (Geng et al.) DAML [Geng et al., 2011], for Domain Adaptation Metric Learning, tackles the general domain adaptation setting. In this case, a classic strategy in DA is to use a term that brings the source and target distribution closer. Following this line of work, Geng et al. regularize the metric using the empirical Maximum Mean Discrepancy [MMD, Gretton et al., 2006], a nonparametric way of measuring the difference in distribution between the source sample

S and the target sample T:

$$MMD(S,T) = \left\| \frac{1}{|S|} \sum_{i=1}^{|S|} \varphi(\boldsymbol{x}_i) - \frac{1}{|T|} \sum_{i=1}^{|T|} \varphi(\boldsymbol{x}'_i) \right\|_{\mathcal{H}}^2,$$

where $\varphi(\boldsymbol{x})$ is a nonlinear feature mapping function that maps $\boldsymbol{x}$ to the Reproducing Kernel Hilbert Space $\mathcal{H}$. The MMD can be computed efficiently using the kernel trick and can thus be used as a (convex) regularizer in kernelized metric learning algorithms (see Section 5.1.1). DAML is thus a trade-off between satisfying the constraints on the labeled source data and finding a projection that minimizes the discrepancy between the source and target distribution. Experiments on face recognition and image annotation tasks in the DA setting highlight the effectiveness of DAML compared to classic metric learning methods.

6.4 HISTOGRAM DATA

We present here three metric learning methods designed specifically for histograms.

χ^2-LMNN (Kedem et al.) Kedem et al. [2012] propose χ^2-LMNN, which is based on the prominent χ^2 histogram distance (see Chapter 2, Eq. 2.2). They propose to generalize this distance with a linear transformation, introducing the following pseudo-distance:

$$\chi_{\boldsymbol{L}}^2(\boldsymbol{x}, \boldsymbol{x}') = \chi^2(\boldsymbol{L}\boldsymbol{x}, \boldsymbol{L}\boldsymbol{x}'),$$

where $\boldsymbol{L} \in \mathbb{R}^{r \times d}$, with the constraint that $\boldsymbol{L}$ preserves the simplex property of the data:

$$\boldsymbol{L} \in \mathcal{P} = \{\boldsymbol{L} \in \mathbb{R}^{r \times d} : \forall \boldsymbol{x} \in \mathcal{S}^d, \boldsymbol{L}\boldsymbol{x} \in \mathcal{S}^r\}.$$

The objective function is the same as LMNN:

$$\min_{\boldsymbol{L} \in \mathcal{P}} \sum_{(\boldsymbol{x}_i, \boldsymbol{x}_j) \in \mathcal{S}_{lmnn}} \chi_{\boldsymbol{L}}^2(\boldsymbol{x}_i, \boldsymbol{x}_j) + \lambda \sum_{(\boldsymbol{x}_i, \boldsymbol{x}_j, \boldsymbol{x}_k) \in \mathcal{R}_{lmnn}} [\gamma + \chi_{\boldsymbol{L}}^2(\boldsymbol{x}_i, \boldsymbol{x}_j) - \chi_{\boldsymbol{L}}^2(\boldsymbol{x}_i, \boldsymbol{x}_k)]_+, \quad (6.7)$$

where $[t]_+ = \max(0, t)$ is the hinge loss and γ the margin parameter.[6] It can be shown that $\boldsymbol{L} \in \mathcal{P}$ if and only if $\boldsymbol{L}$ is element-wise nonnegative and each column sums to 1. This can be enforced by using a change of variable $\boldsymbol{L} = f(\boldsymbol{A})$ where f is the column-wise soft-max operator:

$$f : \mathbb{R}^{r \times d} \to \mathcal{P}, \quad [f(\boldsymbol{A})]_{ij} = \frac{e^{A_{ij}}}{\sum_k e^{A_{kj}}}.$$

The resulting unconstrained problem is solved using a standard subgradient descent procedure. Although (6.7) is nonconvex and thus subject to local optima, experiments show great improvements on histogram data compared to standard histogram metrics and Mahalanobis distance learning methods, as well as promising results for dimensionality reduction (when $r < d$).

[6]In standard LMNN, due to the linearity of the Mahalanobis distance, solutions obtained with different values of the margin only differ up to a scaling factor. The margin can thus be fixed to 1. Here, χ^2 is nonlinear and therefore this value must be tuned.

GML (Cuturi & Avis) While χ^2-LMNN optimizes a simple bin-to-bin histogram distance, Cuturi and Avis [2011] propose to consider the more powerful cross-bin Earth Mover's Distance (EMD) introduced by Rubner et al. [2000], which can be seen as the distance between a source histogram $\boldsymbol{x}$ and a destination histogram $\boldsymbol{x}'$. On an intuitive level, $\boldsymbol{x}$ is viewed as piles of earth at several locations (bins) and $\boldsymbol{x}'$ as several holes, where the value of each feature represents the amount of earth and the capacity of the hole respectively. The EMD is then equal to the minimum amount of effort needed to move all the earth from $\boldsymbol{x}$ to $\boldsymbol{x}'$. The costs of moving one unit of earth from bin i of $\boldsymbol{x}$ to bin j of $\boldsymbol{x}'$ is encoded in the so-called ground distance matrix $\boldsymbol{D} \in \mathbb{R}^{d \times d}$.[7] The computation of EMD amounts to finding the optimal flow matrix $\boldsymbol{F}$, where f_{ij} corresponds to the amount of earth moved from bin i of $\boldsymbol{x}$ to bin j of $\boldsymbol{x}'$. Given the ground distance matrix $\boldsymbol{D}$, $\text{EMD}_{\boldsymbol{D}}(\boldsymbol{x}, \boldsymbol{x}')$ is linear and can be formulated as a linear program:

$$\text{EMD}_{\boldsymbol{D}}(\boldsymbol{x}, \boldsymbol{x}') = \min_{\boldsymbol{f} \in \mathbb{C}(\boldsymbol{x}, \boldsymbol{x}')} \boldsymbol{d}^T \boldsymbol{f},$$

where $\boldsymbol{f}$ and $\boldsymbol{d}$ are respectively the flow and the ground matrices rewritten as vectors for notational simplicity, and $\mathbb{C}(\boldsymbol{x}, \boldsymbol{x}')$ is the convex set of feasible flows (which can be represented as linear constraints).

Ground Metric Learning (GML) aims at learning $\boldsymbol{D}$ based on training histograms $\boldsymbol{x}_1, \dots, \boldsymbol{x}_n$ and side-information $\{w_{ij}\}_{1 \leq i,j \leq n}$ which quantifies how similar $\boldsymbol{x}_i$ and $\boldsymbol{x}_j$ are. We assume $w_{ij} > 0$ for similar pairs and $w_{ij} < 0$ for dissimilar pairs. Given $\boldsymbol{x}_i$, we denote by N_{ik}^+ its k-nearest similar histograms using distance $\text{EMD}_{\boldsymbol{D}}$, and N_{ik}^- is defined likewise as its k-nearest dissimilar histograms. The objective function minimized by GML is

$$f_k(\boldsymbol{D}) = \sum_{i=1}^{n} \sum_{j \in N_{ik}^+ \cup N_{ik}^-} w_{ij} \text{EMD}_{\boldsymbol{D}}(\boldsymbol{x}_i, \boldsymbol{x}_j). \tag{6.8}$$

Problem (6.8) is nonlinear in $\boldsymbol{D}$, and a local minima is found efficiently by a subgradient descent approach. Experiments on image datasets show that GML outperforms standard histogram distances as well as Mahalanobis distance methods.

EMDL (Wang & Guibas) Building on GML and successful Mahalanobis distance learning approaches such as LMNN, Wang and Guibas [2012] aim at learning the ground matrix of EMD in the more flexible setting where the algorithm is given a set of relative constraints $\mathcal{R}$ that must be satisfied with a large margin. The problem is formulated as

$$\begin{aligned} \min_{\boldsymbol{D} \in \mathbb{D}} \quad & \|\boldsymbol{D}\|_{\mathcal{F}}^2 + C \sum_{i,j,k} \xi_{ijk} \\ \text{s.t.} \quad & \text{EMD}_{\boldsymbol{D}}(\boldsymbol{x}_i, \boldsymbol{x}_k) - \text{EMD}_{\boldsymbol{D}}(\boldsymbol{x}_i, \boldsymbol{x}_j) \geq 1 - \xi_{ijk} \quad \forall (\boldsymbol{x}_i, \boldsymbol{x}_j, \boldsymbol{x}_k) \in \mathcal{R}, \end{aligned} \tag{6.9}$$

[7] For EMD to be proper distance, $\boldsymbol{D}$ must satisfy the following $\forall i, j, k \in \{1, \dots, d\}$: (i) $d_{ij} \geq 0$, (ii) $d_{ii} = 0$, (iii) $d_{ij} = d_{ji}$ and (iv) $d_{ij} \leq d_{ik} + d_{kj}$.

where $\mathbb{D} = \left\{ \boldsymbol{D} \in \mathbb{R}^{d \times d} : \forall i, j \in \{1, \ldots, d\}, d_{ij} \geq 0, d_{ii} = 0 \right\}$ and $C \geq 0$ is the trade-off parameter.[8] The authors also propose a pair-based formulation. Problem (6.9) is bi-convex and is solved using an alternating procedure: first fix the ground metric and solve for the flow matrices (this amounts to a set of standard EMD problems), then solve for the ground matrix given the flows (this is a quadratic program). The algorithm stops when the changes in the ground matrix are sufficiently small. The procedure is subject to local optima (because (6.9) is not jointly convex) and is not guaranteed to converge: there is a need for a trade-off parameter α between stable but conservative updates (i.e., staying close to the previous ground matrix) and aggressive but less stable updates. Experiments on face verification datasets confirm that EMDL improves upon standard histogram distances and Mahalanobis distance learning methods, but no comparison to GML is provided.

[8]Note that unlike in GML, $\boldsymbol{D} \in \mathbb{D}$ may not be a valid distance matrix. In this case, $\text{EMD}_{\boldsymbol{D}}$ is not a proper distance.

CHAPTER 7

Metric Learning for Structured Data

In many application domains, data naturally come structured, as opposed to the "flat" feature vector representation we have focused on so far. Indeed, instances can be in the form of strings, such as words, text documents or DNA sequences; trees like XML documents, secondary structure of RNA or parse trees; and graphs, such as networks, 3D objects or molecules. In the context of structured data, metrics are especially appealing because they can be used as a proxy to access data without having to directly manipulate these complex objects. Many metrics used for structured data actually rely on representing structured objects as feature vectors, such as some string kernels [see Lodhi et al., 2002, and variants] or bags-of-(visual)-words [Li and Perona, 2005, Salton et al., 1975]. In this case, metric learning can simply be performed on the feature vector representation, but this strategy can imply a significant loss of structural information.

On the other hand, there exist metrics that operate directly on the structured objects and can thus capture more structural distortions. However, learning such metrics is challenging because most of structured metrics are combinatorial by nature, which explains why it has received less attention than metric learning for feature vectors. In this chapter, we mainly focus on edit distances, which basically measure (in terms of number of operations) the cost of turning an object into another (see Section 2.2.2). Edit distances have attracted most of the interest in the context of metric learning for structured data because (i) they are variants for a variety of objects (sequences, trees and graphs), and (ii) they are naturally amenable to learning due to their parameterization by a cost matrix.

This chapter is organized as follows. We review string edit distance learning in Section 7.1, while methods for trees and graphs are covered in Section 7.2. We end this chapter with a short overview of metric learning for time series (Section 7.3).

7.1 STRING EDIT DISTANCE LEARNING

The standard string edit distance, often called Levenshtein edit distance, is based on a unit cost for all operations. However, this might not reflect the reality of the considered task: for example, in typographical error correction, the probability that a user hits the Q key instead of W on a QWERTY keyboard is much higher than the probability that he/she hits Q instead of Y. For some applications, such as protein alignment or handwritten digit recognition, hand-tuned cost

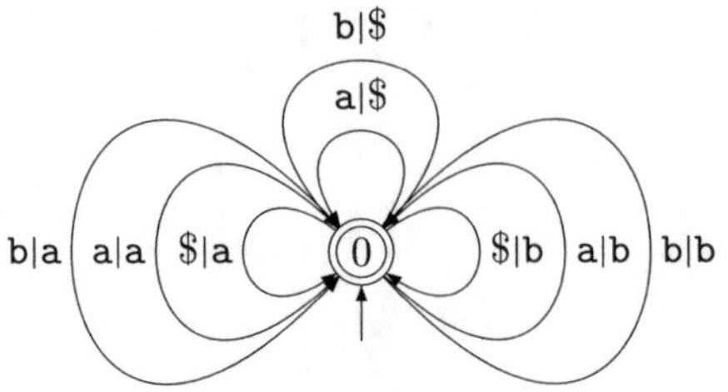

Figure 7.1: A memoryless stochastic transducer that models the edit probability of any pair of strings built from $\Sigma = \{\mathtt{a}, \mathtt{b}\}$. Edit probabilities assigned to each transition are not shown here for the sake of readability.

matrices may be available [Dayhoff et al., 1978, Henikoff and Henikoff, 1992, Micó and Oncina, 1998].

In this section, we review methods to automatically learn the cost matrix $\boldsymbol{C}$ of the string edit distance.

7.1.1 PROBABILISTIC METHODS

Optimizing the edit distance is challenging because the optimal sequence of operations depends on the edit costs themselves, and therefore updating the costs may change the optimal edit script. Most general-purpose approaches get around this problem by considering a stochastic variant of the edit distance, where the cost matrix defines a probability distribution over the edit operations. One can then define an edit similarity as the posterior probability $p_e(\mathsf{x}'|\mathsf{x})$ that an input string x is turned into an output string x'. This corresponds to summing over all possible edit scripts that turn x into x' instead of only considering the optimal script. Such a stochastic edit process can be represented as a probabilistic model, such as a stochastic transducer (Figure 7.1), and one can estimate the parameters of the model (i.e., the cost matrix $\boldsymbol{C}$) that maximize the expected log-likelihood of positive pairs. This is typically done via an EM-like iterative procedure [Dempster et al., 1977]. Note that unlike the standard edit distance, the obtained edit similarity does not usually satisfy the properties of a distance (in fact, it is often not symmetric and rarely satisfies the triangular inequality).

Ristad and Yianilos The first method for learning a string edit metric, in the form of a generative model, was proposed by Ristad and Yianilos [1998].[1] They use a memoryless stochastic transducer which models the joint probability of a pair $p_e(\mathsf{x}, \mathsf{x}')$ from which $p_e(\mathsf{x}'|\mathsf{x})$ can be estimated. Parameter estimation is performed with an EM procedure. The Expectation step takes the form of a probabilistic version of the dynamic programing algorithm of the standard edit distance. The M-step aims at maximizing the likelihood of the training pairs of strings built from

[1]An implementation is available within the SEDiL platform [Boyer et al., 2008]: `http://labh-curien.univ-st-etienne.fr/SEDiL/`.

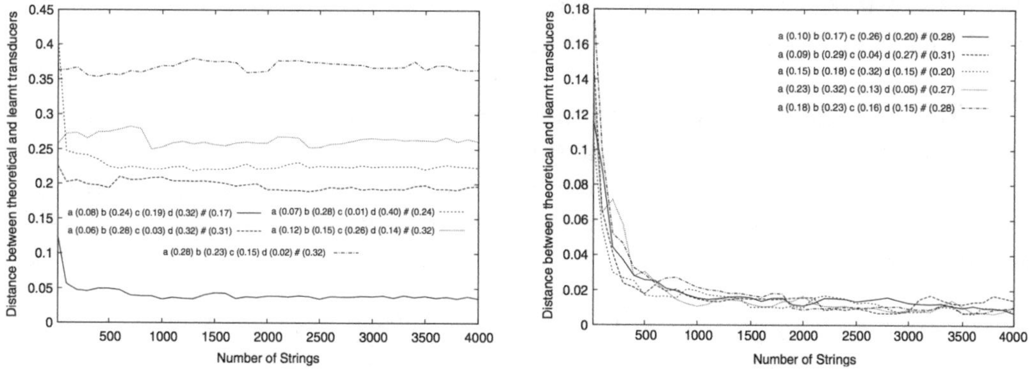

Figure 7.2: Comparison between generative and discriminative edit distance models. On the left, Ristadt and Yianilos's model which learns a joint distribution induces a bias on the input distribution, while Oncina and Sebban's model (on the right) overcomes this shortcoming by learning a conditional distribution.

an alphabet χ so as to define a joint distribution over the edit operations:

$$\sum_{(u,v)\in(\chi\cup\{\$\})^2\setminus\{\$,\$\}} C_{uv} + c(\#) = 1, \quad \text{with } c(\#) > 0 \text{ and } C_{uv} \quad 0,$$

where \$ is the empty symbol, # is a termination symbol and $c(\#)$ the associated cost (probability).

Note that Bilenko and Mooney [2003] extended this approach to the Needleman-Wunsch score with affine gap penalty and applied it to duplicate detection. To deal with the tendency of Maximum Likelihood estimators to overfit when the number of parameters is large (in this case, when the alphabet size is large), Takasu [2009] proposes a Bayesian parameter estimation of pair-HMM providing a way to smooth the estimation.

Oncina and Sebban The work of Oncina and Sebban [2006] describes three levels of bias induced by the use of generative models: (i) dependence between edit operations, (ii) dependence between the costs and the prior distribution of strings $p_e(\mathsf{x})$, and (iii) the fact that to obtain the posterior probability one must divide by the empirical estimate of $p_e(\mathsf{x})$. These biases are highlighted by empirical experiments conducted with the method of Ristad and Yianilos [1998] and that are reported in Figure 7.2 (on the left). The authors simulate a memoryless transducer representing a target joint distribution (matrix) C_{uv} over the edit operations $(u, v) \in (\{a, b, c, d,\} \cup \{\$\})^2 \setminus \{\$, \$\}$. From C_{uv}, the marginal target input distribution $C_{u.}$ can be deduced as follows: $C_{u.} = \sum_v C_{uv}$, that corresponds to the sum of the edit probabilities over the output alphabet. Generating a growing set of input strings according to different random marginal distributions $C_{u.}$, the authors show that the only way to approximately recover the joint target transducer (i.e., C_{uv}) is to set $C_{u.} = C_{u.}$ (solid line on the Figure). Said differently, this means that the stochastic distance learned by Ristad and Yianilos depends on the a priori distribution of the input strings, which is undesirable.

To address these limitations, Oncina and Sebban propose the use of a conditional transducer as a discriminative model that directly encodes the posterior probability $p(\mathsf{x}'|\mathsf{x})$ that an input string x is turned into an output string x' using edit operations. Parameter estimation is also done with EM where the maximization step differs from that of Ristad and Yianilos [1998] as shown below:

$$\forall u \in \Sigma, \sum_{v \in \Sigma \cup \{\$\}} C_{v|u} + \sum_{v \in \Sigma} C_{v|\$} = 1, \quad \text{with} \sum_{v \in \Sigma} C_{v|\$} + c(\#) = 1.$$

Figure 7.2 (on the right) shows that using this discriminative model allows one to overcome the drawback of Ristad and Yianilos's model. Regardless of the marginal input distribution, the learned conditional transducer converges toward the target one.

Note that in order to allow the use of negative pairs, McCallum et al. [2005] consider another discriminative model, conditional random fields, that can deal with positive and negative pairs in specific states, still using EM for parameter estimation.

7.1.2 GRADIENT DESCENT METHODS

The use of EM has two main drawbacks: (i) it may converge to a local optimum, and (ii) distance calculations must be done at each iteration, which can be very costly if the size of the alphabet and/or the length of the strings are large. The following methods get around one or both of these drawbacks by formulating the learning problem in the form of an optimization problem that can be efficiently solved by a gradient descent procedure.

Saigo et al. Saigo et al. [2006] manage to avoid the need for an iterative procedure like EM in the context of detecting remote homology in protein sequences.[2] They learn the parameters of the Smith-Waterman score which is plugged in their local alignment kernel k_{LA} where all the possible local alignments π for changing x into x' are taken into account [Saigo et al., 2004]:

$$k_{LA}(\mathsf{x}, \mathsf{x}') = \sum_{\pi} e^{t \cdot s(\mathsf{x}, \mathsf{x}', \pi)}. \tag{7.1}$$

In the above formula, t is a parameter and $s(\mathsf{x}, \mathsf{x}', \pi)$ is the corresponding score of π and defined as follows:

$$s(\mathsf{x}, \mathsf{x}', \pi) = \sum_{u,v \in \Sigma} n_{u,v}(\mathsf{x}, \mathsf{x}', \pi) \cdot C_{uv} - n_{g_d}(\mathsf{x}, \mathsf{x}', \pi) \cdot g_d - n_{g_e}(\mathsf{x}, \mathsf{x}', \pi) \cdot g_e, \tag{7.2}$$

[2]Source code available at: `http://sunflower.kuicr.kyoto-u.ac.jp/~hiroto/project/optaa.html`.

where $n_{u,v}(\mathsf{x}, \mathsf{x}', \pi)$ is the number of times that symbol u is aligned with v while g_d and g_e, along with their corresponding number of occurrences $n_{g_d}(\mathsf{x}, \mathsf{x}', \pi)$ and $n_{g_e}(\mathsf{x}, \mathsf{x}', \pi)$, are two parameters dealing respectively with the opening and extension of gaps.

Unlike the Smith-Waterman score, k_{LA} is differentiable and can be optimized by a gradient descent procedure. The objective function that they optimize is meant to favor the discrimination between positive and negative examples, but this is done by only using positive pairs of distant homologs. The approach has two additional drawbacks: (i) the objective function is nonconvex and thus subject to local minima, and (ii) in general, k_{LA} does not fulfill the properties of a kernel.

GESL (Bellet et al.) Bellet et al. [2011, 2012a] propose a convex programming approach to learn edit similarity functions from both positive and negative pairs without requiring a costly iterative procedure.[3] They use the following simplified edit function:

$$e_C(\mathsf{x}, \mathsf{x}') = \sum_{(u,v)\in(\Sigma\cup\{\$\})^2\setminus\{\$,\$\}} C_{uv} \cdot \#_{uv}(\mathsf{x}, \mathsf{x}'),$$

where $\#_{uv}(\mathsf{x}, \mathsf{x}')$ is the number of times the operation $u \rightarrow v$ appears in the Levenshtein script. Therefore, e_C can be optimized directly since the sequence of operations is fixed (it does not depend on the costs). The authors optimize the nonlinear similarity $K_C(\mathsf{x}, \mathsf{x}') = 2\exp(-e_C(\mathsf{x}, \mathsf{x}')) - 1$, derived from e_C. Note that K_C is not required to be PSD nor symmetric. GESL (Good Edit Similarity Learning) is expressed as follows:

$$\begin{aligned} \min_{C,B_1,B_2} \quad & \frac{1}{n^2}\sum_{z_i,z_j} \ell(C, z_i, z_j) \quad + \quad \beta\|C\|_{\mathcal{F}}^2 \\ \text{s.t.} \quad & B_1 \geq -\log(\tfrac{1}{2}), \quad 0 \leq B_2 \leq -\log(\tfrac{1}{2}), \quad B_1 - B_2 = \eta_\gamma, \end{aligned}$$

where $\beta \geq 0$ is a regularization parameter, $\eta_\gamma \geq 0$ a parameter corresponding to a desired "margin" and

$$\ell(C, z_i, z_j) = \begin{cases} [B_1 - e_C(\mathsf{x}_i, \mathsf{x}_j)]_+ \text{ if } y_i \neq y_j \\ [e_C(\mathsf{x}_i, \mathsf{x}_j) - B_2]_+ \text{ if } y_i = y_j. \end{cases}$$

GESL essentially learns the edit cost matrix C so as to optimize the (ϵ, γ, τ)-goodness [Balcan et al., 2008a] of the similarity $K_C(\mathsf{x}, \mathsf{x}')$ and thereby enjoys generalization guarantees both for the learned similarity and for the resulting linear classifier (see Chapter 8). A potential drawback of GESL is that it optimized a simplified variant of the edit distance, although this does not seem to be an issue in practice. Note that GESL can be straightforwardly adapted to learn tree or graph edit similarities [Bellet et al., 2012a].

[3]Source code available at: `http://www-bcf.usc.edu/~bellet/`.

7.2 TREE AND GRAPH EDIT DISTANCE LEARNING

In this section, we briefly review the main approaches in tree/graph edit distance learning. We do not delve into the details of these approaches as they are essentially adaptations of stochastic string edit distance learning presented in Section 7.1.1.

Bernard et al. Extending the work of Ristad and Yianilos [1998] and Oncina and Sebban [2006] on string edit similarity learning, Bernard et al. [2006, 2008] propose both a generative and a discriminative model for learning tree edit costs. They rely on the tree edit distance by Selkow [1977]—which is cheaper to compute than that of Zhang and Shasha [1989]—and adapt the updates of EM to this case.

Boyer et al. The work of Boyer et al. [2007] tackles the more complex variant of the tree edit distance [Zhang and Shasha, 1989], which allows the insertion and deletion of single nodes instead of entire subtrees only. Parameter estimation in the generative model is also based on EM.

Dalvi et al. The work of Dalvi et al. [2009] points out a limitation of the approach of Bernard et al. [2006, 2008]: they model a distribution over tree edit scripts rather than over the trees themselves, and unlike the case of strings, there is no bijection between the edit scripts and the trees. Recovering the correct conditional probability with respect to trees requires a careful and costly procedure. They propose a more complex conditional transducer that models the conditional probability over trees and use again EM for parameter estimation.

Emms The work of Emms [2012] points out a theoretical limitation of the approach of Boyer et al. [2007]: the authors use a factorization that turns out to be incorrect in some cases. Emms shows that a correct factorization exists when only considering the edit script of highest probability instead of all possible scripts, and derives the corresponding EM updates. An obvious drawback is that the output of the model is not the probability $p(\mathsf{x}'|\mathsf{x})$. Moreover, the approach is prone to overfitting and requires smoothing and other heuristics (such as a final step of zeroing-out the diagonal of the cost matrix).

Neuhaus and Bunke In their paper, Neuhaus and Bunke [2007] learn a (more general) graph edit similarity, where each edit operation is modeled by a Gaussian mixture density. Parameter estimation is done using an EM-like algorithm. Unfortunately, the approach is intractable: the complexity of the EM procedure is exponential in the number of nodes (and so is the computation of the distance).

7.3 METRIC LEARNING FOR TIME SERIES

Temporal sequences are another kind of structured data where the structuring depends on time information. In this context, dynamic time warping (DTW) [Kruskall and Liberman, 1983] is the most commonly used method for aligning two time series x and x'. It consists in warping the time axis so as to find the shortest path to go from x to x' (see Figure 7.3).

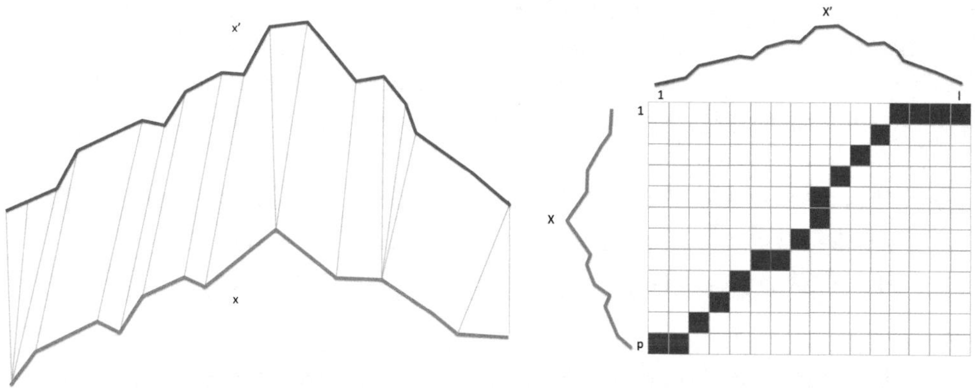

Figure 7.3: On the left: two time series that are optimally aligned by warping the time axis. On the right, the optimal alignment corresponding to the shortest path to change the sequence x into x'.

As done for the edit distance, DTW optimization has received great interest during the past few years to capture the specificities of the tasks at hand. In order to enhance its behavior to deal with classification (mostly in nearest-neighbor classification) or clustering tasks, efforts have been made to constrain time warping to avoid pathological alignment, to handle dynamics or to focus on local events rather than the entire time series values. Such an optimization is achieved by assessing the parameters of the measure, namely the warping constraints, the time weighting, or the underlying cost function. Since the pioneering work of Sakoe [1978], several methods have been developed in the literature, generally based on multiple bands global path estimation [Ratanamahatana and Keogh, 2004, Yu et al., 2011], genetic algorithms [Gaudin and Nicoloyannis, 2006], warping-based penalization [Jeong et al., 2010] or adaptive weighted (linear) cost functions [Wu et al., 2009].

Note that most of the previous approaches assume that (i) the time series of a given class share a similar global behavior and (ii) two different classes do not partly share similar shapes. To overcome this limitation, Frambourg et al. [2013] present a multiple temporal matching approach which uses a criterion based on the variance/covariance to reweight matched observations according to the induced variability within and between classes.

MLTSA (Lajugie et al.) Very recently, Lajugie et al. [2014b] cast the problem of optimizing multivariate time series alignment as a Mahalanobis distance learning problem and proposed MLTSA (Metric Learning for Temporal Sequence Alignment). Let $X = (A, B)$ be a pair of multivariate time series of same dimension p but of possibly different lengths T_A and T_B, where $A \in \mathbb{R}^{T_A \times p}$, $B \in \mathbb{R}^{T_B \times p}$ and rows of A (resp. B) are denoted by $a_1, ..., a_{T_A}$ (resp. $b_1, ..., b_{T_B}$). The pairwise affinity matrix $C(X, W)_{i,j}$, parameterized by a PSD matrix W, is given as follows:

$$C(X, W)_{i,j} = -(a_i - b_j)^T W(a_i - b_j).$$

Using groundtruth in the form of a set S of n training pairs of time series $(X^i, Y^i)_{i=1}^n$ for which the true alignment Y^i is known, they optimize an area loss to find W which allows the minimization over S of the area between Y^i and the predited alignment parameterized by W. An approximation of the area loss between two alignment matrices Y_1 and Y_2 is defined as follows:

$$
\begin{aligned}
l(Y_1, Y_2) &= \frac{1}{2}[Tr(Y_1^T(L_{T_A}^T L_{T_A} - D_{T_A})Y_1) + Tr(D_{T_A} Y_1 \mathbf{1}_{T_B} \mathbf{1}_{T_A}^T) + Tr(L_{T_A} Y_2 \mathbf{1}_{T_B} \mathbf{1}_{T_A}^T) \\
&- 2Tr(Y_2^T(L_{T_A}^T L_{T_A} - D_{T_A})Y_1) + Tr(Y_1(L_{T_B} L_{T_B}^T - D_{T_B})Y_2) \\
&+ Tr(Y_1 D_{T_B} \mathbf{1}_{T_B} \mathbf{1}_{T_A}^T) + Tr(Y_2^T L_{T_B} L_{T_B}^T Y_2) - 2Tr(Y_2 L_{T_B} L_{T_B}^T Y_1^T)],
\end{aligned}
$$

where $D_T = \lambda_{max}(L_T^T L_T) I_{T \times T}$ with $\lambda_{max}(U)$ the largest eigenvalue of U. Note that $l(Y_1, Y_2)$ is concave. The authors plug it in a margin-rescaled problem which is solved efficiently using a Frank-Wolfe algorithm. Experiments are conducted on audio signal alignment, for which the proposed method leads to improvements in the alignment performance.

CHAPTER 8

Generalization Guarantees for Metric Learning

The derivation of guarantees on the generalization performance of the learned model is a wide topic in statistical learning theory [Valiant, 1984, Vapnik and Chervonenkis, 1971]. Assuming that data points are independent and identically distributed (i.i.d.) according to some (unknown but fixed) distribution μ, one essentially aims at bounding the deviation of the *true risk* of the learned model (its performance on unseen data) from its *empirical risk* (its performance on the training sample). This deviation is typically a function of the number of training examples, and some notion of complexity of the model such as the VC dimension [Vapnik and Chervonenkis, 1971], the fat-shattering dimension [Alon et al., 1997] or the Rademacher complexity [Bartlett and Mendelson, 2002, Koltchinskii, 2001].

In the context of metric learning, we claim that the question of generalization can be seen as two-fold [Bellet, 2012], as illustrated in Figure 8.1:

- The *consistency* of the learned metric: the goal is to bound the deviation between its performance on training pair/triplet constraints and its performance on unseen constraints.

- The *prediction performance* of the learned metric: when it is used to improve a prediction model (e.g., k-NN or a linear classifier), the goal is to characterize the generalization performance of this predictor with respect to the learned metric.

We begin this chapter with a high-level overview of existing results in Section 8.1, before going through some of the key results in more detail. Section 8.2 is dedicated to the derivation of consistency bounds for metric, with a focus on the frameworks of uniform stability and algorithmic robustness. In Section 8.3, we show how to obtain generalization guarantees for a linear classifier based on a learned metric.

8.1 OVERVIEW OF EXISTING WORK

In the literature, the body of work that investigates generalization guarantees for metric learning can be grouped into three categories.

Consistency bounds for batch methods Given a training sample $\mathcal{T} = \{z_i = (\boldsymbol{x}_i, y_i)\}_{i=1}^n$ drawn i.i.d. from an unknown distribution μ, let us consider fully supervised Mahalanobis metric learn-

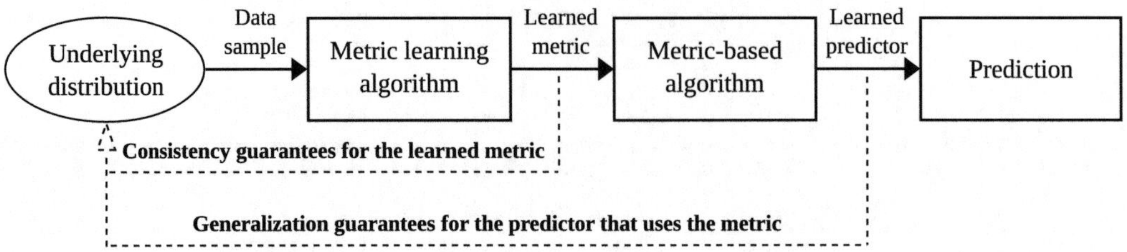

Figure 8.1: The two-fold problem of generalization in metric learning. We may be interested in the generalization ability of the learned metric itself: can we say anything about its consistency on unseen data drawn from the same distribution? Furthermore, we may also be interested in the generalization ability of the predictor using that metric: can we relate its performance on unseen data to the quality of the learned metric?

ing of the following general form:

$$\min_{M \in \mathbb{S}_+^d} \quad \frac{1}{n^2} \sum_{z_i, z_j \in \mathcal{T}} \ell(d_M^2, z_i, z_j) \quad + \quad \lambda R(M),$$

where $R(M)$ is the regularizer, λ the regularization parameter and the loss function ℓ is of the form $\ell(d_M^2, z_i, z_j) = g(y_i y_j [c - d_M^2(\boldsymbol{x}_i, \boldsymbol{x}_j)])$ with $c > 0$ a decision threshold variable and g convex and Lipschitz continuous. The loss function $g(\cdot)$ should output a small value when its input is large positive and a large value when it is large negative. This includes popular loss functions such as the hinge loss.

Some recent work has proposed to study the convergence of the empirical risk (as measured by ℓ on pairs from $\mathcal{T}$) to the true risk over the unknown probability distribution μ. The framework proposed by Bian and Tao [2011, 2012] is somewhat restrictive as it relies on strong assumptions on the moments of μ and cannot accommodate any regularization (a constraint to bound M is used instead). Jin et al. [2009] introduce a notion of uniform stability [Bousquet and Elisseeff, 2002] adapted to the case of metric learning (where training data consists of pairs) to derive generalization bounds for the case of Frobenius norm regularization. Bellet and Habrard [2015] demonstrate how to adapt the more flexible notion of algorithmic robustness [Xu and Mannor, 2012] to the metric learning setting to derive (loose) generalization bounds for any matrix norm (including sparse-inducing norms) as regularizer. They also show that a weak notion of robustness is necessary and sufficient for metric learning algorithms to generalize well. Lastly, Cao et al. [2012a] propose a notion of Rademacher complexity [Bartlett and Mendelson, 2002] dependent on the regularizer to derive tighter bounds for several matrix norms. All these results can be easily adapted to other (non-Mahalanobis) linear metric learning formulations.

Regret bound conversion for online methods Wang et al. [2012d, 2013b] point out that existing proof techniques to convert regret bounds into generalization bounds [see for instance Cesa-Bianchi and Gentile, 2008] do not hold for pairwise loss functions such as those used in metric learning. They derive an alternative framework for dealing with a pairwise loss where at each round, the online algorithm receives a new instance and must pair it with all previously seen data points. As this is expensive or even infeasible in practice, Kar et al. [2013] propose to use a buffer containing only a bounded number of the most recent instances. They are also able to obtain tighter bounds based on a notion of Rademacher complexity, essentially adapting and extending the work of Cao et al. [2012a].

Classification performance of a learned metric To the best of our knowledge, the question of the prediction performance of the learned metric has only been addressed in the context of metric learning for linear classification. Bellet et al. [2011, 2012a,b] rely on the theory of learning with (ϵ, γ, τ)-good similarity functions [Balcan et al., 2008a], which makes the link between properties of a similarity function and the generalization of a linear classifier built from this similarity. Bellet et al. propose to use (ϵ, γ, τ)-goodness as the objective function for metric learning, and show that it is possible to derive consistency bounds for the learned metric that imply generalization guarantees for the linear classifier. Guo and Ying [2014] extend these results to several matrix norms using a Rademacher complexity analysis, based on techniques from Cao et al. [2012a].

In the rest of this chapter, we focus on two frameworks to show the consistency of batch metric learning approaches in Section 8.2. Then, in Section 8.3, we show how to derive guarantees for a linear classifier from the consistency guarantees of a learned similarity according to two lines of work: we first present the approach of Bellet et al. [2012b] and in a second step we provide a quick overview of the results obtained by Guo and Ying [2014].

8.2 CONSISTENCY BOUNDS FOR METRIC LEARNING

This section is devoted to the following question: what can we say about the performance of the learned metric on unseen constraints? An obvious strategy to obtain generalization bounds would be to apply existing results for classic supervised learning (where training data consist of individual labeled instances) by considering each pair or triplet as an i.i.d. sample. However, the i.i.d. assumption is violated in metric learning since the training pairs/triplets are constructed from the training sample. For this reason, establishing generalization guarantees for the learned metric is challenging and only recently has this question been investigated from a theoretical standpoint.

8.2.1 DEFINITIONS

Our goal throughout this section will be to bound the generalization error of a metric. This is formalized by the following definition.

Definition 8.1 Let μ be a distribution over $\mathcal{Z} = \mathcal{X} \times \mathcal{Y}$. The *generalization error* of a metric $d : \mathcal{X} \times \mathcal{X} \to \mathbb{R}$ with respect to a pair-based loss function ℓ is the expected loss suffered by d

over pairs of examples drawn from the distribution μ:

$$\epsilon(d) = \mathbb{E}_{z,z'\sim\mu}\ell(d, z, z').$$

Bounding the generalization error is generally achieved by bounding its deviation to the empirical error measured on the training sample.

Definition 8.2 Let $\mathcal{T} = \{z_i = (\boldsymbol{x}_i, y_i)\}_{i=1}^n \in \mathcal{Z}^n$ be a training sample drawn i.i.d. from some distribution μ. The *empirical error* of a metric d with respect to a loss function ℓ over $\mathcal{T}$ is the average loss suffered by d on the pairs constructed from $\mathcal{T}$:

$$\epsilon_{\mathcal{T}}(d) = \frac{1}{n^2} \sum_{z_i, z_j \in \mathcal{T}} \ell(d, z_i, z_j).$$

The generalization bounds we will derive are known as PAC (Probably Approximately Correct) bounds [Valiant, 1984] and have the following form:

$$P(|\epsilon(d) - \epsilon_{\mathcal{T}}(d)| < \eta) > 1 - \delta,$$

where $\eta \geq 0$ and $0 \leq \delta \leq 1$. To obtain a consistency result, we need η to depend on the size of the learning sample $\mathcal{T}$ such that it converges to 0 when $\mathcal{T}$ goes to infinity.

In the rest of this section, we show how to obtain such bounds. For simplicity, we assume that the metric is a squared Mahalanobis distance d_M^2 (the analysis straightforwardly extends to other metrics). Section 8.2.2 introduces uniform stability analysis, which is restricted to particular forms of regularizers. Section 8.2.3 focuses on algorithmic robustness, a property that can be used to derive consistency guarantees for a larger class of metric learning formulations but gives looser bounds. Note that an approach based on a notion of Rademacher complexity for metric learning will be presented later in Section 8.3.2.

8.2.2 BOUNDS BASED ON UNIFORM STABILITY

Uniform stability [Bousquet and Elisseeff, 2002] is a prominent framework in statistical learning theory to derive generalization guarantees. We present an adaptation of this framework to metric learning due to Jin et al. [2009] and show how to use it to obtain a generalization bound for a standard metric learning formulation.

Preliminaries

We first introduce technical notions that are needed for the analysis. We assume that the loss function ℓ is k-Lipschitz (Definition 8.3) with respect to its first argument (which corresponds to the parameter $\boldsymbol{M}$ of the metric) and is (σ, m)-admissible (Definition 8.4).

Definition 8.3 A loss function $\ell(d^2_{\boldsymbol{M}}, z_1, z_2)$ is k*-Lipschitz* with respect to its first argument if for any parameter matrices $\boldsymbol{M}, \boldsymbol{M}'$ and any pair of labeled examples (z_1, z_2):

$$|\ell(d^2_{\boldsymbol{M}}, z_1, z_2) - \ell(d^2_{\boldsymbol{M}'}, z_1, z_2)| \leq k\|\boldsymbol{M} - \boldsymbol{M}'\|_{\mathcal{F}}.$$

Definition 8.4 A loss function $\ell(d^2_{\boldsymbol{M}}, z_1, z_2)$ is (σ, m)*-admissible*, with respect to $d^2_{\boldsymbol{M}}$, if (i) it is convex with respect to its first argument and (ii) the following condition holds:

$$\forall z_1, z_2, z_3, z_4, \quad |\ell(d^2_{\boldsymbol{M}}, z_1, z_2) - \ell(d^2_{\boldsymbol{M}}, z_3, z_4)| \leq \sigma|y_{12} - y_{34}| + m$$

where $z_i = (\boldsymbol{x}_i, y_i)$, and $y_{ij} = 1$ if $y_i = y_j$ and -1 otherwise $(i, j = 1, 2, 3, 4)$.

Definition 8.4 requires that the deviation of the losses between two pairs of examples be bounded by a value that depends only on the labels and some constants independent from the examples and $\boldsymbol{M}$. It follows that the labels must be bounded, which is not a strong assumption in the classification setting we are interesting in. In our case, we have binary labels $y_{ij} \in \{-1, 1\}$ implying that the quantity $|y_{12} - y_{34}|$ is either 0 or 2.

In this section, we assume the regularizer $R(\boldsymbol{M})$ to be strictly convex and for simplicity we set it to be the Frobenius norm $R(\boldsymbol{M}) = \|\boldsymbol{M}\|^2_{\mathcal{F}}$. We consider the following objective function for Mahalanobis distance learning:

$$\begin{aligned} F_{\mathcal{T}} &= \frac{1}{n^2} \sum_{z_i, z_j \in \mathcal{T}} \ell(d^2_{\boldsymbol{M}}, z_i, z_j) + \lambda\|\boldsymbol{M}\|_{\mathcal{F}} \\ &= \epsilon_{\mathcal{T}}(d^2_{\boldsymbol{M}}) + \lambda\|\boldsymbol{M}\|_{\mathcal{F}}. \end{aligned}$$

From the convexity of ℓ with respect to its first argument (the parameter matrix $\boldsymbol{M}$) it follows that $F_{\mathcal{T}}$, ϵ and $\epsilon_{\mathcal{T}}$ are convex. The associated optimization problem is as follows:

$$\min_{\boldsymbol{M} \in \mathbb{S}^d_+} F_{\mathcal{T}}. \tag{8.1}$$

In the following, we denote by $\boldsymbol{M}_{\mathcal{T}}$ the parameter matrix learned from a sample $\mathcal{T}$.

Uniform Stability and Generalization Guarantees

We present the notion of uniform stability and derive a generalization bound for the metric learning formation given in (8.1). Roughly speaking, a learning algorithm is stable [Bousquet and Elisseeff, 2002] if its output does not change significantly under a small modification of the training sample. We consider the following definition of uniform stability, stating that replacing one example must lead to a variation bounded in $O(1/n)$ in terms of infinite norm.

Definition 8.5 A learning algorithm has a *uniform stability* in $\frac{\kappa}{n}$, where κ is a positive constant, if

$$\forall(\mathcal{T}, z), \forall i, \sup_{z_1,z_2} |\ell(d^2_{M_\mathcal{T}}, z_1, z_2) - \ell(d^2_{M_{\mathcal{T}^{i,z}}}, z_1, z_2)| \leq \frac{\kappa}{n},$$

where $\mathcal{T}^{i,z}$ is the training sample obtained by replacing $z_i \in \mathcal{T}$ by a new example z.

To prove the property of uniform stability of formulation (8.1), we need the following lemma related to the associated regularizer and the k-Lipschitz property of ℓ.

Lemma 8.6 *Let $F_\mathcal{T}$ and $F_{\mathcal{T}^{i,z}}$ be the functions to optimize, $\boldsymbol{M}_\mathcal{T}$ and $\boldsymbol{M}_{\mathcal{T}^{i,z}}$ their corresponding minimizers, and λ the regularization parameter. Let $\Delta\boldsymbol{M} = (\boldsymbol{M}_\mathcal{T} - \boldsymbol{M}_{\mathcal{T}^{i,z}})$. For any $t \in [0, 1]$:*

$$\|\boldsymbol{M}_\mathcal{T}\|^2_\mathcal{F} - \|\boldsymbol{M}_\mathcal{T} - t\Delta\boldsymbol{M}\|^2_\mathcal{F} + \|\boldsymbol{M}_{\mathcal{T}^{i,z}}\|^2_\mathcal{F} - \|\boldsymbol{M}_{\mathcal{T}^{i,z}} + t\Delta\boldsymbol{M}\|^2_\mathcal{F} \leq \frac{4kt}{\lambda n}\|\Delta\boldsymbol{M}\|_\mathcal{F}.$$

The proof is given in Appendix A.1.1. We can now prove the uniform stability property for formulation (8.1).

Theorem 8.7 *Let $\mathcal{T}$ be a learning sample of n labeled examples, an algorithm solving* (8.1) *has a uniform stability in $\frac{\kappa}{n}$, where $\kappa = \frac{4k^2}{\lambda}$.* The proof is given in Appendix A.1.2.

Now, using the property of stability, we can derive our generalization bound for $\epsilon(d^2_{M_\mathcal{T}})$. This is done by using the McDiarmid inequality [McDiarmid, 1989].

Theorem 8.8 *Let $X_1, \ldots, X_n$ be n independent random variables taking values in X and let $Z = f(X_1, \ldots, X_n)$. If for each $1 \leq i \leq n$, there exists a constant c_i such that*

$$\sup_{x_1,\ldots,x_n,x'_i \in \mathcal{X}} |f(x_1, \ldots, x_n) - f(x_1, \ldots, x'_i, \ldots, x_n)| \leq c_i, \forall 1 \leq i \leq n,$$

$$\textit{then for any } \beta > 0, \qquad \Pr[|Z - \mathbf{E}[Z]| \geq \beta] \leq 2\exp\left(\frac{-2\beta^2}{\sum_{i=1}^n c_i^2}\right).$$

To derive our bound on $\epsilon(d^2_{M_\mathcal{T}})$, we consider the quantity $D_\mathcal{T} = \epsilon(d^2_{M_\mathcal{T}}) - \epsilon_\mathcal{T}(d^2_{M_\mathcal{T}})$, and simply need to replace Z by $D_\mathcal{T}$ in Theorem 8.8 and to bound $\boldsymbol{E}_\mathcal{T}[D_\mathcal{T}]$ and $|D_\mathcal{T} - D_{\mathcal{T}^{i,z}}|$, which is shown by the following lemmas. The proofs can be found in Appendix A.1.3 and Appendix A.1.4.

Lemma 8.9 *For any learning method of estimation error $D_\mathcal{T}$ and satisfying a uniform stability in $\frac{\kappa}{n}$, we have $\boldsymbol{E}_\mathcal{T}[D_\mathcal{T}] \leq \frac{2\kappa}{n}$.*

Lemma 8.10 *For any parameter matrix $\boldsymbol{M}$ using n training examples, and any loss function ℓ satisfying (σ, m)-admissibility, we have the following bound:*

$$\forall i, 1 \leq i \leq n, \quad \forall \boldsymbol{z}, \quad |D_{\mathcal{T}} - D_{\mathcal{T}^{i,z}}| \leq \frac{2\kappa}{n} + \frac{2(2\sigma + m)}{n}.$$

We are now able to derive our generalization bound for $\epsilon(d^2_{\boldsymbol{M}_{\mathcal{T}}})$.

Theorem 8.11 Generalization bound with uniform stability. *Let $\mathcal{T}$ be a sample of n randomly selected training examples and let $\boldsymbol{M}_{\mathcal{T}}$ be the parameter matrix learned from Problem (8.1) with stability $\frac{\kappa}{n}$. Assuming that $\ell(d^2_{\boldsymbol{M}_{\mathcal{T}}}, \boldsymbol{z}, \boldsymbol{z}')$ is k-Lipschitz and (σ, m)-admissible, with probability $1 - \delta$, we have the following bound for $\epsilon(d^2_{\boldsymbol{M}_{\mathcal{T}}})$:*

$$\epsilon(d^2_{\boldsymbol{M}_{\mathcal{T}}}) \leq \epsilon_{\mathcal{T}}(d^2_{\boldsymbol{M}_{\mathcal{T}}}) + 2\frac{\kappa}{n} + (2\kappa + 2\,(2\sigma + m))\sqrt{\frac{\ln(2/\delta)}{2n}}$$

with $\kappa = \frac{4k^2}{\lambda}$.

Proof. Recall that $D_{\mathcal{T}} = \epsilon(d^2_{\boldsymbol{M}_{\mathcal{T}}}) - \epsilon_{\mathcal{T}}(d^2_{\boldsymbol{M}_{\mathcal{T}}})$. From Lemma 8.10, we get

$$|D_{\mathcal{T}} - D_{\mathcal{T}^{i,z}}| \leq \sup_{\mathcal{T}, z'} |D_{\mathcal{T}} - D_{\mathcal{T}^{i,z'}}| \leq \frac{2\kappa + B}{n} \text{ with } B = 2(2\sigma + m).$$

Then by applying the McDiarmid inequality, we have

$$\Pr[|D_{\mathcal{T}} - \boldsymbol{E}_{\mathcal{T}}[D_{\mathcal{T}}]| \geq \beta] \leq 2\exp\left(-\frac{2\beta^2}{\sum_{i=1}^{n}\frac{(2\kappa+B)^2}{n^2}}\right) \leq 2\exp\left(-\frac{2\beta^2}{\frac{(2\kappa+B)^2}{n}}\right). \tag{8.2}$$

By fixing $\delta = 2\exp\left(-\frac{2\beta^2}{(2\kappa+B)^2/n}\right)$, we get $\beta = (2\kappa + B)\sqrt{\frac{\ln(2/\delta)}{2n}}$. Finally, from (8.2), Lemma 8.9 and the definition of $D_{\mathcal{T}}$, we have with probability at least $1 - \delta$:

$$D_{\mathcal{T}} < \boldsymbol{E}_{\mathcal{T}}[D_{\mathcal{T}}] + \beta \Rightarrow \epsilon(d^2_{\boldsymbol{M}_{\mathcal{T}}}) < \epsilon_{\mathcal{T}}(d^2_{\boldsymbol{M}_{\mathcal{T}}}) + 2\frac{\kappa}{n} + (2\kappa + B)\sqrt{\frac{\ln(2/\delta)}{2n}}.$$

□

Application to the hinge loss

We consider the case where the loss function is the hinge loss: $\ell(d^2_{\boldsymbol{M}}, z_i, z_j) = \left[y_{ij}[c - d^2_{\boldsymbol{M}}(\boldsymbol{x}_i, \boldsymbol{x}_j)]\right]_+$ where $y_{ij} = 1$ if $y_i = y_j$ and -1 otherwise. To derive a bound for this specific case, we need to find the constants of k-lipschitzness and (σ, m)-admissibility related to ℓ and derive a bound on the norm of $\boldsymbol{M}_{\mathcal{T}}$. This is done in the following lemmas (the proofs can be found in Appendix A.1.5, Appendix A.1.6 and Appendix A.1.7).

Lemma 8.12 *Let $\boldsymbol{M}_{\mathcal{T}}$ be an optimal solution of Problem (8.1), then*

$$\|\boldsymbol{M}_{\mathcal{T}}\|_{\mathcal{F}} \leq \frac{c}{\lambda}.$$

Lemma 8.13 *The loss function $\ell(d_M^2, z_i, z_j) = \left[y_{ij}[c - d_M^2(\boldsymbol{x}_i, \boldsymbol{x}_j)]\right]_+$ is k-lipschitz with $k = 4B^2$.*

Lemma 8.14 *The loss function $\ell(d_M^2, z_i, z_j) = \left[y_{ij}[c - d_M^2(\boldsymbol{x}_i, \boldsymbol{x}_j)]\right]_+$ is (σ, m)-admissible with $\sigma = c$ and $m = 8B^2 \frac{c}{\lambda}$.*

Setting $k = 4B^2$, $\sigma = c$ and $m = 8B^2 \frac{c}{\lambda}$ in Theorem 8.11 gives us the generalization bound for the hinge loss case. Applying the same process to other loss functions allows one to derive similar guarantees.

Uniform stability is a convenient tool to derive generalization guarantees for strictly convex metric learning formulations, but it cannot deal with sparsity inducing norms. Indeed, it is known that sparse algorithms are not stable [Xu et al., 2012], and thus stability-based analysis fails to assess the generalization ability of sparse metric learning approaches presented earlier in this book [such as McFee and Lanckriet, 2010, Rosales and Fung, 2006, Shi et al., 2014, Ying et al., 2009]. In the next section, we present generalization guarantees obtained with algorithmic robustness, which can accommodate a larger class of metric learning algorithms.

8.2.3 BOUNDS BASED ON ALGORITHMIC ROBUSTNESS

In this section, we study the generalization ability of metric learning algorithms according to a notion of algorithmic robustness. This framework, introduced by Xu and Mannor [2010, 2012], allows one to derive generalization bounds when the variation in the loss associated with two *nearby* training and testing examples is bounded. The notion of closeness relies on a partition of the input space into different regions such that two examples in the same region are considered close. Unlike uniform stability (Section 8.2.2), which is based on a notion of proximity of two training sets, robustness is thus based on geometric proximity of individual training examples.

Robustness has been successfully used to derive generalization bounds in the classic supervised learning setting, with results for SVM, LASSO, etc. We present here an extension of the framework of robustness algorithmic to metric learning due to Bellet and Habrard [2015] and give the associated generalization result. Then, we illustrate the wide applicability of the framework by deriving generalization bounds for a family of problems with various regularizers.

Robustness and Generalization for Metric Learning

Let $\mathcal{T} = \{z_i = (\boldsymbol{x}_i, y_i)\}_{i=1}^n \in \mathcal{Z}^n$ be a training sample drawn i.i.d. from an unknown distribution μ. We denote by $\mathcal{P}(\mathcal{T})$ the set of all possible pairs that can be built from $\mathcal{T}$: $\mathcal{P}(\mathcal{T}) = \{(z_1, z_1), \ldots, (z_1, z_n), \ldots, (z_n, z_n)\}$.

We assume that $\mathcal{X}$ is a compact convex metric space with respect to a norm $\|\cdot\|$ such that $\mathcal{X} \subset \mathbb{R}^d$, thus there exists a constant R such that $\forall \boldsymbol{x} \in \mathcal{X}$, $\|\boldsymbol{x}\| \leq R$. When considering metric spaces, the partition of $\mathcal{Z}$, that we will use in this section, can be obtained by the notion of covering number [Kolmogorov and Tikhomirov, 1961].

Definition 8.15 For a metric space $(\mathcal{X}, \rho)$, and $T \subset \mathcal{X}$, we say that $\hat{T} \subset T$ is a γ-cover of T, if $\forall t \in T, \exists \hat{t} \in \hat{T}$ such that $\rho(t, t') \leq \gamma$. The γ-covering number of T is

$$\mathcal{N}(\gamma, T, \rho) = \min\{|\hat{T}| : \hat{T} \text{ is a } \gamma\text{-cover of } T\}.$$

When $\mathcal{X}$ is a compact convex space, for any $\gamma > 0$, the quantity $\mathcal{N}(\gamma, X, \rho)$ is finite, leading to a finite cover. If we consider the space $\mathcal{Z}$, note that the label set can be partitioned into $|Y|$ sets. Thus, $\mathcal{Z}$ can be partitioned into $|Y|\mathcal{N}(\gamma, X, \rho)$ subsets, i.e., we consider a partition over $\mathcal{X}$ for each class in Y, such that if two instances $z_1 = (\boldsymbol{x}_1, y_1)$, $z_2 = (\boldsymbol{x}_2, y_2)$ belong to the same subset, then $y_1 = y_1$ and $\rho(\boldsymbol{x}_1, \boldsymbol{x}_2) \leq \gamma$.

We now present the notion of robustness for metric learning. The original formulation for pointwise loss functions [Xu and Mannor, 2012] is based on the deviation between the loss associated with two training and testing examples that are "close." The idea of the adaptation to metric learning is to use the partition of $\mathcal{Z}$ at the pair level: if a test pair is close to a training pair, then the losses associated with each pair must be similar. Two pairs are close when each instance of the first pair fall into the same subset of the partition of $\mathcal{Z}$ as the corresponding instance of the other pair, as shown in Figure 8.2. A metric learning algorithm with this property is said to be robust. This notion is formalized as follows.

Definition 8.16 A metric learning algorithm $\mathcal{A}$ is $(K, \omega(\cdot))$-*robust* for $K \in \mathbb{N}$ and $\omega(\cdot) : (\mathcal{Z} \times \mathcal{Z})^n \to \mathbb{R}$ if $\mathcal{Z}$ can be partitioned into K disjoints sets, denoted by $\{C_i\}_{i=1}^K$, such that for all sample $\mathcal{T} \in \mathcal{Z}^n$ and the pair set $\mathcal{P}(\mathcal{T})$ associated to this sample, the following holds:
$\forall (z_1, z_2) \in \mathcal{P}(\mathcal{T}), \forall z_1', z_2' \in \mathcal{Z}, \forall i, j = 1, \ldots, K$: if $z_1, z_1' \in C_i$ and $z_2, z_2' \in C_j$ then

$$|\ell(d^2_{M_{\mathcal{P}(\mathcal{T})}}, z_1, z_2) - \ell(d^2_{M_{\mathcal{P}(\mathcal{T})}}, z_1', z_2')| \leq \omega(\mathcal{P}(\mathcal{T})). \tag{8.3}$$

K and $\omega(\cdot)$ quantify the robustness of the algorithm and depend on the training sample. The property of robustness is required for every training pair of the sample; we will later see that this property can be relaxed.

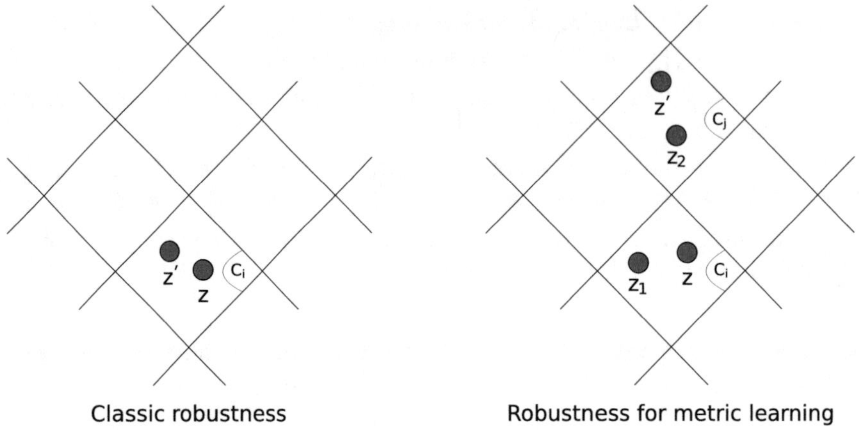

Figure 8.2: Illustration of the robustness property in the classic and metric learning settings. In this example, we use a cover based on the L_1 norm. In the classic definition, if any example z' falls in the same region C_i as a training example z, then the deviation between their loss must be bounded. In the metric learning definition proposed in this work, for any pair (z, z') and a training pair (z_1, z_2), if z, z_1 belong to some region C_i and z', z_2 to some region C_j, then the deviation between the loss of these two pairs must be bounded.

Note that this definition of robustness can be easily extended to triplet-based metric learning algorithms. Instead of considering all the pairs $\mathcal{P}(\mathcal{T})$ from an i.i.d. sample $\mathcal{T}$, we take the admissible triplet set $tri\mathcal{P}(\mathcal{T})$ of $\mathcal{T}$ such that $(z_1, z_2, z_3) \in tri\mathcal{P}(\mathcal{T})$ means z_1 and z_2 share the same label while z_1 and z_3 have different ones, with the interpretation that z_1 must be more similar to z_2 than to z_3. The robustness property can then be expressed by: $\forall (z_1, z_2, z_3) \in tri\mathcal{P}(\mathcal{T}), \forall z_1', z_2', z_3' \in \mathcal{Z}, \forall i, j, k = 1, \ldots, K$: if $z_1, z_1' \in C_i$, $z_2, z_2' \in C_j$ and $z_3, z_3' \in C_k$ then

$$|\ell(d^2_{tri\mathcal{P}(\mathcal{T})}, z_1, z_2, z_3) - \ell(d^2_{tri\mathcal{P}(\mathcal{T})}, z_1', z_2', z_3')| \leq \omega(tri\mathcal{P}(\mathcal{T})). \tag{8.4}$$

Generalization Guarantees of Robust Algorithms

We now give a generalization bound for metric learning algorithms satisfying the property of robustness (Definition 8.16).

Theorem 8.17 Generalization bound with algorithmic robustness. *If a learning algorithm $\mathcal{A}$ is $(K, \omega(\cdot))$-robust and the training sample consists of the pairs $\mathcal{P}(\mathcal{T})$ obtained from a sample $\mathcal{T}$ generated by n i.i.d. draws from μ, then for any $\delta > 0$, with probability at least $1 - \delta$ we have:*

$$|\epsilon(d^2_{M_\mathcal{T}}) - \epsilon_\mathcal{T}(d^2_{M_\mathcal{T}})| \leq \omega(\mathcal{P}(\mathcal{T})) + 2B\sqrt{\frac{2K \ln 2 + 2 \ln 1/\delta}{n}}.$$

The proof is given in Appendix A.2.2.

The previous bound depends on $\omega(\mathcal{P}(\mathcal{T}))$ and K, which depend on the cover chosen for $\mathcal{Z}$. The former goes to zero as K increases, ensuring the validity of the bound.[1] If for any K, the associated $\omega(\cdot)$ is a constant (i.e., $\omega_K(\mathcal{T}) = \omega_K$) for any $\mathcal{T}$, we can prove a bound that holds uniformly for all K:

$$|\epsilon(d^2_{M_\mathcal{T}}) - \epsilon_\mathcal{T}(d^2_{M_\mathcal{T}})| \leq \inf_{K \geq 1} \left[\omega_K + 2B\sqrt{\frac{2K\ln 2 + 2\ln 1/\delta}{n}} \right].$$

This also gives an insight into the objective of any robust algorithm: according to a partition of the labeled input space, given two regions, minimize the maximum loss over pairs of examples belonging to each region.

For triplet-based metric learning algorithms, by following the definition of robustness given by Equation 8.4 and adapting straightforwardly the losses to triplets such that they output zero for nonadmissible ones, Theorem 8.17 can be easily extended to obtain the following generalization bound:

$$|\epsilon(d^2_{M_\mathcal{T}}) - \epsilon_\mathcal{T}(d^2_{M_\mathcal{T}})| \leq \omega(tri\mathcal{P}(\mathcal{T})) + 3B\sqrt{\frac{2K\ln 2 + 2\ln 1/\delta}{n}}. \tag{8.5}$$

Pseudo-robustness

In the analysis above, we require the robustness property to be satisfied for every training pair. It is possible to relax this requirement such that it must hold only for a subset of the possible pairs and still obtain generalization guarantees, as shown below.

Definition 8.18 An algorithm $\mathcal{A}$ is $(K, \omega(\cdot), \hat{p}_n(\cdot))$ *pseudo-robust* for $K \in \mathbb{N}$, $\omega(\cdot) : (\mathcal{Z} \times \mathcal{Z})^n \to \mathbb{R}$ and $\hat{p}_n(\cdot) : (\mathcal{Z} \times \mathcal{Z})^n \to \{1, \ldots, n^2\}$, if $\mathcal{Z}$ can be partitioned into K disjoints sets, denoted by $\{C_i\}_{i=1}^K$, such that for all $\mathcal{T} \in \mathcal{Z}^n$ i.i.d. from μ, there exists a subset of training pairs samples $\hat{p}_\mathcal{T} \subseteq \mathcal{P}(\mathcal{T})$, with $|\hat{p}_\mathcal{T}| = \hat{p}_n(\mathcal{P}(\mathcal{T}))$, such that the following holds:
$\forall (z_1, z_2) \in \hat{p}_\mathcal{T}, \forall z_1', z_2' \in \mathcal{Z}, \forall i, j = 1, \ldots, K$: if $z_1, z_1' \in C_i$ and $z_2, z_2' \in C_j$ then

$$|\ell(d^2_{M_\mathcal{T}}, z_1, z_2) - \ell(d^2_{M_\mathcal{T}}, z_1', z_2')| \leq \omega(\mathcal{P}(\mathcal{T})). \tag{8.6}$$

We can easily observe that $(K, \omega(\cdot))$-robust is equivalent to $(K, \omega(\cdot), n^2)$ pseudo-robust. The following theorem gives the generalization guarantees associated with the pseudo-robustness property.

[1] This point will be made clear by the examples provided in Section 8.2.3.

Theorem 8.19 *If a learning algorithm $\mathcal{A}$ is $(K, \omega(\cdot), \hat{p}_n(\cdot))$ pseudo-robust, the training pairs $\mathcal{P}(\mathcal{T})$ come from a sample generated by n i.i.d. draws from μ, then for any $\delta > 0$, with probability at least $1 - \delta$ we have:*

$$|\epsilon(d^2_{M_\mathcal{T}}) - \epsilon_\mathcal{T}(d^2_{M_\mathcal{T}})| \leq \frac{\hat{p}_n(\mathcal{P}(\mathcal{T}))}{n^2}\omega(\mathcal{P}(\mathcal{T})) + B\left(\frac{n^2 - \hat{p}_n(\mathcal{P}(\mathcal{T}))}{n^2} + 2\sqrt{\frac{2K\ln 2 + 2\ln 1/\delta}{n}}\right).$$

The proof is similar to that of Theorem 8.17 and can be found in Bellet and Habrard [2015]. This notion of pseudo-robustness is very relevant to metric learning. Indeed, it is often difficult and potentially damaging to optimize the metric with respect to all possibles pairs, and it has been observed in practice that focusing on a subset of carefully selected pairs (e.g., defined according to nearest-neighbors as in LMNN) gives much better generalization performance. Theorem 8.19 confirms that this principle is well-founded: as long as the robustness property is satisfied for a (large enough) subset of the pairs, the resulting metric has generalization guarantees. Note that this notion of pseudo-robustness can also be easily adapted to triplet-based metric learning.

Examples of Robust Metric Learning Algorithms

We restrict our attention to Mahalanobis distance learning algorithms of the following form:

$$\min_{M \in \mathbb{S}^d_+} \frac{1}{n^2} \sum_{z_i, z_j \in \mathcal{T}} \ell(d^2_M, z_i, z_j) + \lambda\|M\| \tag{8.7}$$

where $\|\cdot\|$ is some matrix norm and $\ell(d^2_M, z_i, z_j) = g(y_{ij}[1 - d^2_M(\boldsymbol{x}_i, \boldsymbol{x}_j)])$ with $y_{ij} = 1$ if $y_i = y_j$ and -1 otherwise. We assume $g(\cdot)$ to be nonnegative and Lipschitz continuous with Lipschitz constant U. Lastly, $g_0 = \sup_{z_i, z_j} g(y_{ij}[1 - d^2_{\mathbf{0}}(\boldsymbol{x}_i, \boldsymbol{x}_j)])$ is the largest loss when $\boldsymbol{M}$ is the null matrix $\mathbf{0}$. The general form (8.7) encompasses many existing metric learning formulations.

To prove the robustness of (8.7), we will need the following theorem, which essentially says that if a metric learning algorithm achieves approximately the same testing loss for testing pairs that are close to each other, then it is robust.

Theorem 8.20 *Fix $\gamma > 0$ and a metric ρ of $\mathcal{Z}$. Suppose $\mathcal{A}$ satisfies:*
$\forall z_1, z_2, z'_1, z'_2 : z_1, z_2 \in \mathcal{T}, \rho(z_1, z'_1) \leq \gamma, \rho(z_2, z'_2) \leq \gamma,$

$$|\ell(d^2_M, z_1, z_2) - \ell(d^2_M, z'_1, z'_2)| \leq \omega(\mathcal{P}(\mathcal{T}))$$

and $\mathcal{N}(\gamma/2, \mathcal{Z}, \rho) < \infty$. Then $\mathcal{A}$ is $(\mathcal{N}(\gamma/2, \mathcal{Z}, \rho), \omega(\mathcal{P}(\mathcal{T})))$-robust.

The proof is given in Appendix A.2.3. This theorem provides a roadmap for deriving generalization guarantees based on the robustness framework. Indeed, given a partition of the input space, one must bound the deviation between the loss for any pair of examples with corresponding elements belonging to the same partitions. This bound is generally a constant that depends on the problem to solve and the thinness of the partition defined by γ. This bound tends to zero as $\gamma \to 0$, which ensures the consistency of the approach.

Following Theorem 8.20, we now prove the robustness of (8.7) when $\|M\|$ is the Frobenius norm.

Example 8.21 Frobenius norm Formulation (8.7) with the Frobenius norm $\|M\| = \|M\|_{\mathcal{F}} = \sqrt{\sum_{i=1}^{d}\sum_{j=1}^{d} m_{ij}^2}$ is $(|Y|\mathcal{N}(\gamma/2, \mathcal{X}, \|\cdot\|_2), \frac{8UR\gamma g_0}{\lambda})$-robust.

Proof. Let M^* be the solution given a training sample $\mathcal{T}$. Thus, due to optimality of M^*, we have

$$\frac{1}{n^2}\sum_{z_i,z_j\in\mathcal{T}} \ell(d_M^2, z_i, z_j) + \lambda\|M\| \le \frac{1}{n^2}\sum_{z_i,z_j\in\mathcal{T}} g(y_{ij}[1 - d_{\mathbf{0}}^2(x_i, x_j)]) + \lambda\|\mathbf{0}\|_{\mathcal{F}} + = g_0,$$

and thus $\|M^*\|_{\mathcal{F}} \le g_0/\lambda$.

We can partition $\mathcal{Z}$ as $|Y|\mathcal{N}(\gamma/2, \mathcal{X}, \|\cdot\|_2)$ sets, such that if z and z' belong to the same set, then $y = y'$ and $\|x - x'\|_2 \le \gamma$. Now, for $z_1, z_2, z_1', z_y'2 \in \mathcal{Z}$, if $y_1 = y_1'$, $\|x_1 - x_1'\|_2 \le \gamma$, $y_2 = y_2'$ and $\|x_2 - x_2'\|_2 \le \gamma$, then:

$$\begin{aligned}
&|g(y_{12}[1 - d_{M^*}^2(x_1, x_2)]) - g(y_{12}'[1 - d_{M^*}^2(x_1', x_2')])| \\
\le\ & U|(x_1 - x_2)^T M^*(x_1 - x_2) - (x_1' - x_2')^T M^*(x_1' - x_2')| \\
=\ & U|(x_1 - x_2)^T M^*(x_1 - x_2) - (x_1 - x_2)^T M^*(x_1' - x_2') \\
& + (x_1 - x_2)^T M^*(x_1' - x_2')| - (x_1' - x_2')^T M^*(x_1' - x_2')| \\
=\ & U|(x_1 - x_2)^T M^*(x_1 - x_2 - (x_1' + x_2')) + \\
& (x_1 - x_2 - (x_1' + x_2'))^T M^*(x_1' + x_2')| \\
\le\ & U(|(x_1 - x_2)^T M^*(x_1 - x_1')| + |(x_1 - x_2)^T M^*(x_2' - x_2)| \\
& + |(x_1 - x_1')^T M^*(x_1' + x_2')| + |(x_2' - x_2)^T M^*(x_1' + x_2')|) \\
\le\ & U(\|x_1 - x_2\|_2\|M^*\|_{\mathcal{F}}\|x_1 - x_1'\|_2 + \|x_1 - x_2\|_2\|M^*\|_{\mathcal{F}}\|x_2' - x_2\|_2 \\
& + \|x_1 - x_1'\|_2\|M^*\|_{\mathcal{F}}\|x_1' - x_2'\|_2 + \|x_2' - x_2\|_2\|M^*\|_{\mathcal{F}}\|x_1' - x_2'\|_2) \\
\le\ & \frac{8UR\gamma g_0}{\lambda}.
\end{aligned}$$

Hence, the example holds by Theorem 8.20. □

The key advantage of robustness over stability is that it can accommodate arbitrary p-norms (or even any regularizer which is bounded below by some p-norm), using equivalence of norms arguments. To illustrate this, we show the robustness when $\|M\|$ is the L_1 norm [used in Rosales and Fung, 2006] which promotes sparsity at the entry level, the $L_{2,1}$ norm [used e.g., in Shi et al., 2014, Ying et al., 2009] which induces sparsity at the column/row level, and the trace norm [used e.g., in McFee and Lanckriet, 2010] which favors low-rank matrices. The proofs are reminiscent of that of Example 8.21 and can be found in A.2.4 and A.2.5, respectively.

Example 8.22 L_1 **norm** Formulation (8.7) with $\|M\| = \|M\|_1$ is $(|Y|\mathcal{N}(\gamma, \mathcal{X}, \|\cdot\|_1), \frac{8UR\gamma g_0}{c})$-robust.

Example 8.23 **$L_{2,1}$ norm and trace norm** Consider formulation (8.7) with $\|\boldsymbol{M}\| = \|\boldsymbol{M}\|_{2,1} = \sum_{i=1}^{d} \|m^i\|_2$, where m^i is the i-th column of $\boldsymbol{M}$. This formulation is $(|Y|\mathcal{N}(\gamma, \mathcal{X}, \|\cdot\|_2), \frac{8UR\gamma g_0}{\lambda})$-robust. The same holds with the trace norm $\|\boldsymbol{M}\|_*$, which is the sum of the singular values of $\boldsymbol{M}$.

We can also derive guarantees for kernelized metric learning (see Section 5.1.1).

Example 8.24 **Kernelization** Consider the kernelized version of (8.7):

$$\min_{\boldsymbol{M} \in \mathbb{S}_+^d} \frac{1}{n^2} \sum_{(s_i, s_j) \in p_s} g(y_{ij}[1 - d_{\boldsymbol{M}}^2(\phi(\boldsymbol{x}_i), \phi(\boldsymbol{x}_j))]) + \lambda \|\boldsymbol{M}\|_{\mathbb{H}}, \tag{8.8}$$

where $\phi(\cdot)$ is a feature mapping to a kernel space $\mathbb{H}$, $\|\cdot\|_{\mathbb{H}}$ the norm function of $\mathbb{H}$ and $k(\cdot,\cdot)$ the kernel function. Consider a cover of $\mathcal{X}$ by the metric $\|\cdot\|_2$ ($\mathcal{X}$ being compact) and let $f_{\mathbb{H}}(\gamma) \triangleq \max_{\boldsymbol{M}, \boldsymbol{b} \in \mathcal{X}, \|\boldsymbol{M}-\boldsymbol{b}\|_2 \le \gamma} (k(\boldsymbol{M}, \boldsymbol{M}) + k(\boldsymbol{b}, \boldsymbol{b}) - 2k(\boldsymbol{M}, \boldsymbol{b}))$ and $B_\gamma = \max_{\boldsymbol{x} \in \mathcal{X}} \sqrt{k(\boldsymbol{x}, \boldsymbol{x})}$. If the kernel function is continuous, B_γ and $f_{\mathbb{H}}$ are finite for any $\gamma > 0$ and thus Algorithm 8.8 is $(|Y|\mathcal{N}(\gamma, \mathcal{X}, \|\cdot\|_2), \frac{8UB_\gamma \sqrt{f_{\mathbb{H}}} g_0}{\lambda})$-robust.
The proof is given in Appendix A.2.6.

Example 8.25 We can easily prove similar results for other forms of metrics using the same technique. For instance, when the function is a bilinear similarity $S_{\boldsymbol{M}}(\boldsymbol{x}_i, \boldsymbol{x}_j) = \boldsymbol{x}_i^T \boldsymbol{M} \boldsymbol{x}_j$ where $\boldsymbol{M}$ is usually not constrained to be PSD [Bellet et al., 2012b, Chechik et al., 2009, Qamar et al., 2008], we can improve the robustness to $2UR\gamma g_0/\lambda$.
The proof follows the same principle of the proof provided for Example 8.21.

Example 8.26 Using triplet-based robustness (Equation 8.4), we can for instance show the robustness of two popular triplet-based metric learning approaches [Schultz and Joachims, 2003, Ying et al., 2009] for which no generalization guarantees were known (to the best of our knowledge). These algorithms have the following form:

$$\min_{\boldsymbol{M} \in \mathbb{S}_+^d} \frac{1}{|trip\mathcal{P}(\mathcal{T})|} \sum_{(\boldsymbol{z}_i, \boldsymbol{z}_j, \boldsymbol{z}_k) \in trip\mathcal{P}(\mathcal{T})} [1 - (\boldsymbol{x}_i - \boldsymbol{x}_k)^T \boldsymbol{M} (\boldsymbol{x}_i - \boldsymbol{x}_k) + (\boldsymbol{x}_i - \boldsymbol{x}_j)^T \boldsymbol{M} (\boldsymbol{x}_i - \boldsymbol{x}_j)]_+ + \lambda \|\boldsymbol{M}\|,$$

where $\|\boldsymbol{M}\| = \|\boldsymbol{M}\|_{\mathcal{F}}$ in Schultz and Joachims [2003] and $\|\boldsymbol{M}\| = \|\boldsymbol{M}\|_{1,2}$ in Ying et al. [2009]. These methods are $(\mathcal{N}(\gamma, \mathcal{Z}, \|\cdot\|_2), \frac{16UR\gamma g_0}{\lambda})$-robust.

The proof uses the same techniques as in Examples 8.21 and 8.23, the additional factor 2 comes from the use of triplets instead of pairs.

Note that the price to pay for this general framework is the relative looseness of the resulting bounds, since they rely on a constants (related to covering numbers) that can be large and/or hard to estimate.

8.3 GUARANTEES ON CLASSIFICATION PERFORMANCE

Consistency bounds for the learned metric, such as those obtained in the previous section, do not directly translate into generalization guarantees on the classification performance. The missing link is a relationship between the quality of a metric and its performance in classification. In this section, we present some recent work that managed to design metric learning approaches with guarantees on the accuracy of a linear classifier based on the learned metric.

8.3.1 GOOD SIMILARITY LEARNING FOR LINEAR CLASSIFICATION

In this section, we first present the theory of (ϵ, γ, τ)-good similarity functions, which makes the link between some properties of a similarity function and the performance of a linear classifier based on this similarity. Then we show how this framework is leveraged in Bellet et al. [2012b] to design a similarity learning approach with guarantees on the classification performance.

Good Similarity Function for Linear Classification

In this section, we review the theory of (ϵ, γ, τ)-good similarity functions [Balcan and Blum, 2006, Balcan et al., 2008a], which is motivated by two limitations of the notion of good kernel: (i) a similarity function must be symmetric and PSD to be a valid kernel, which is often violated by natural measures of similarity, and (ii) the notion of good kernel is defined according to an implicit, possibly unknown projection space, making it hard to design appropriate kernels for a given application. To address these drawbacks, they propose the following notion of good similarity function.

Definition 8.27 A similarity function $S : \mathcal{X} \times \mathcal{X} \rightarrow [-1, 1]$ is an (ϵ, γ, τ)-*good similarity function* for a binary classification problem following a distribution μ if there exists a (random) indicator function $R(\boldsymbol{x})$ defining a (probabilistic) set of "reasonable points" such that the following conditions hold:

1. A $1 - \epsilon$ probability mass of examples $(\boldsymbol{x}, y)$ satisfy

$$\boldsymbol{E}_{(\boldsymbol{x}', y') \sim \mu}[y y' S(\boldsymbol{x}, \boldsymbol{x}') | R(\boldsymbol{x}')] \geq \gamma.$$

2. $\Pr_{\boldsymbol{x}'}[R(\boldsymbol{x}')] \geq \tau$.

The first condition is essentially requiring that *a* $1 - \epsilon$ *proportion of examples x are on average* 2γ *more similar to random reasonable examples of the same class than to random reasonable examples of the opposite class* and the second condition that *at least a* τ *proportion of the examples are reasonable.* Note that the reasonable points may be given by some domain knowledge (they can be viewed as prototypes) or are automatically selected from a set of so-called landmarks by solving a simple linear problem which is essentially a 1-norm SVM problem (see Balcan et al. [2008a] for details).

Definition 8.27 is very interesting in two respects. First, it includes all good kernels as well as some non-PSD similarity functions. In that sense, this is a strict generalization of the notion

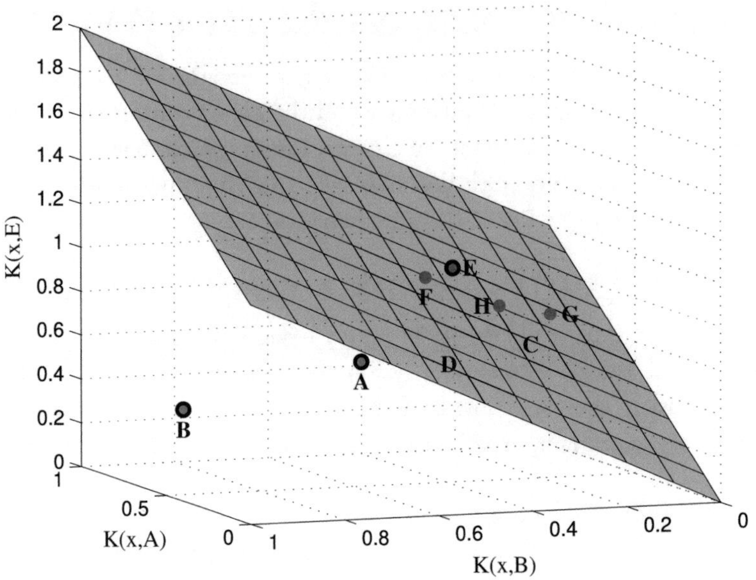

Figure 8.3: Linear separator $\boldsymbol{\alpha}$ learned from a training set $\{(A,+1),(B,+1),(C,+1),(D,+1),(E,-1),(F,-1),(G,-1),(H,-1)\}$ based on a similarity function S defined as the opposite value of the normalized Euclidean distance. A, B and E are reasonable points. The linear separator is learned in the 3D-space of the similarities to these reasonable points with respect to K.

of good kernel [Balcan et al., 2008a]. Second, these conditions are sufficient to learn a good linear space, i.e., to induce a low-error linear separator $\boldsymbol{\alpha}$ in the space of the similarities to the reasonable points: $h(\cdot) = \sum_{(\boldsymbol{x}',y')\sim\mu|R(\boldsymbol{x}')} y'S(\cdot,\boldsymbol{x}')$, as illustrated in Figure 8.3. This is formalized by the following theorem.

Theorem 8.28 *Let S be an (ϵ,γ,τ)-good similarity function in hinge loss for a binary classification problem following a distribution μ. For any $\epsilon_1 > 0$ and $0 \leq \delta \leq \gamma\epsilon_1/4$, let $L = \{\boldsymbol{x}'_1,\ldots,\boldsymbol{x}'_{d_{land}}\}$ be a (potentially unlabeled) sample of $d_{land} = \frac{2}{\tau}\left(\log(2/\delta) + 16\frac{\log(2/\delta)}{(\epsilon_1\gamma)^2}\right)$ landmarks drawn from μ. Consider the mapping $\phi^L : \mathbb{R}^d \rightarrow \mathbb{R}^{d_{land}}$ defined as follows: $\phi^L_i(\boldsymbol{x}) = S(\boldsymbol{x},\boldsymbol{x}'_i)$, $i \in \{1,\ldots,d_{land}\}$. Then, with probability at least $1-\delta$ over the random sample L, the induced distribution $\phi^L(\mu)$ in $\mathbb{R}^{d_{land}}$ has a linear separator $\boldsymbol{\alpha}$ of error at most $\epsilon + \epsilon_1$ at margin γ.*

Therefore, if we are given an (ϵ,γ,τ)-good similarity function for a learning problem P and enough points, then with high probability there exists a low-error linear separator $\boldsymbol{\alpha}$ in the explicit "ϕ-space," which is essentially the space of the similarities to the d_{land} "landmarks." As Balcan et al. mention, using d_u (potentially unlabeled) landmark examples and d_l labeled examples, we can efficiently find this separator $\boldsymbol{\alpha} \in \mathbb{R}^{d_u}$ by solving the following linear program (LP):[2]

$$\min_{\boldsymbol{\alpha}} \sum_{i=1}^{d_l} \left[1 - \sum_{j=1}^{d_u} \alpha_j y_i S(\boldsymbol{x}_i, \boldsymbol{x}'_j) \right]_+ + \lambda \|\boldsymbol{\alpha}\|_1. \tag{8.9}$$

Note that Problem (8.9) is essentially an L_1-regularized linear SVM [Zhu et al., 2003] with an empirical similarity map [Balcan and Blum, 2006], and can be efficiently solved. The L_1-regularization induces sparsity (zero coordinates) in $\boldsymbol{\alpha}$, which speeds up predictions. One can control the amount of sparsity by using the parameter λ (the larger λ, the sparser $\boldsymbol{\alpha}$).

To sum up, the performance of the linear classifier theoretically depends on how well the similarity function satisfies Definition 8.27. However, for real-world problems, it may be poorly satisfied by standard similarity functions. In the case of learned metric, generalization guarantees proved in Section 8.2 could be used to obtain a loose upper bound on the ϵ error rate of Definition 8.27, since they focus on the error over any pair of examples while (ϵ, γ, τ)-goodness requires only an average error over reasonable points. Another strategy is to use (ϵ, γ, τ)-goodness as a well-founded objective function for similarity learning to obtain a direct bound on the generalization error of the linear classifier. This is the approach taken in the next section.

Generalization Bound for the Linear Classifier

Formulation We consider the bilinear similarity $S_{\boldsymbol{M}} : \mathbb{R}^d \times \mathbb{R}^d \to \mathbb{R}$:

$$S_{\boldsymbol{M}}(\boldsymbol{x}, \boldsymbol{x}') = \boldsymbol{x}^T \boldsymbol{M} \boldsymbol{x}',$$

parameterized by the matrix $\boldsymbol{M} \in \mathbb{R}^{d \times d}$, which is not constrained to be PSD nor symmetric. In order to satisfy the condition $S_{\boldsymbol{M}} \in [-1, 1]$ in Definition 8.27, we assume that inputs are normalized such that $||\boldsymbol{x}||_2 \leq 1$, and we require $||\boldsymbol{M}||_{\mathcal{F}} \leq 1$.

To learn $\boldsymbol{M}$ so as to optimize the (ϵ, γ, τ)-goodness of $S_{\boldsymbol{M}}$, we consider a training sample of n labeled points $\mathcal{T} = \{\boldsymbol{z}_i = (\boldsymbol{x}_i, y_i)\}_{i=1}^n$ and a sample of n_R labeled reasonable points $R = \{\boldsymbol{z}_k = (\boldsymbol{x}_k, y_k)\}_{k=1}^{n_R}$. In practice, R is a subset of $\mathcal{T}$ with $n_R = \hat{\tau} n$ ($\hat{\tau} \in \,]0, 1]$). In the absence of background knowledge, it can be drawn randomly or according to a given criterion (e.g., diversity [Kar and Jain, 2011]). The formulation, called SLLC for Similarity Learning for Linear Classification Bellet et al. [2012b], is given by

$$\begin{aligned} \min_{\boldsymbol{M} \in \mathbb{R}^{d \times d}} \quad & \frac{1}{n} \sum_{i=1}^{n} \ell(\boldsymbol{M}, \boldsymbol{z}_i, R) \quad + \quad \lambda \|\boldsymbol{M}\|_{\mathcal{F}}^2 \\ \text{s.t.} \quad & \ell(\boldsymbol{M}, \boldsymbol{z}_i, R) = [1 - y_i \frac{1}{\gamma n_R} \sum_{k=1}^{n_R} y_k S_{\boldsymbol{M}}(\boldsymbol{x}_i, \boldsymbol{x}_k)]_+, \end{aligned} \tag{8.10}$$

[2]The original formulation proposed by Balcan et al. [2008a] was actually L_1-constrained. We turned it into an equivalent L_1-regularized one.

where γ is the margin and λ the regularization parameter. Note that SLLC follows directly from Definition 8.27: given a margin γ and a set of reasonable points R, it minimizes the amount of margin violations ϵ over $\mathcal{T}$ with respect to R. Therefore, $\hat{\epsilon} = \frac{1}{n}\sum_{i=1}^{n} \ell(\boldsymbol{M}, \boldsymbol{z}_i, R)$ and $\hat{\tau}$ are empirical estimates for ϵ and τ respectively.

Preliminaries We now want to derive a generalization bound that guarantees the consistency of SLLC and thus the (ϵ, γ, τ)-goodness in generalization. The similarity $S_{\boldsymbol{M}}$ is optimized according to a set R of reasonable points drawn from the training sample. Therefore, these reasonable points may not follow the distribution from which the training sample has been generated. To cope with this non i.i.d. situation, we derive a generalization bound according to the framework of uniform stability Bousquet and Elisseeff [2002] previously used in Section 8.2.2, but here the objective is to derive guarantees for the linear classifier using the learned metric. For this reason, we consider the stability property with respect to the pointwise loss instead of the pairwise loss.

Definition 8.29 A learning algorithm has a *uniform stability* in $\frac{\kappa}{n}$ w.r.t. a loss function ℓ, with κ a positive constant, if

$$\forall \mathcal{T}, |\mathcal{T}| = n, \forall i, 1 \leq i \leq n, \sup_{\boldsymbol{z}} |\ell(\boldsymbol{M}_{\mathcal{T}}, \boldsymbol{z}) - \ell(\boldsymbol{M}_{\mathcal{T}^i}, \boldsymbol{z})| \leq \frac{\kappa}{n},$$

where $\boldsymbol{M}_{\mathcal{T}}$ is the model learned from the sample $\mathcal{T}$, $\boldsymbol{M}_{\mathcal{T}^i}$ the model learned from the sample $\mathcal{T}^i$, $\mathcal{T}^i$ is obtained from $\mathcal{T}$ by replacing the ith example $\boldsymbol{z}_i \in \mathcal{T}$ by another example $\boldsymbol{z}_i'$ independent from $\mathcal{T}$ and drawn from μ. $\ell(\boldsymbol{M}, \boldsymbol{z})$ is the loss for an example $\boldsymbol{z}$.

When this definition is fulfilled, Bousquet and Elisseeff [2002] have shown that the following generalization bound holds.

Theorem 8.30 *Let $\delta > 0$ and $n > 1$. For any algorithm with uniform stability κ/n using a loss function bounded by 1, with probability $1-\delta$ over the random draw of $\mathcal{T}$:*

$$\epsilon(\boldsymbol{M}_{\mathcal{T}}) < \epsilon_{\mathcal{T}}(\boldsymbol{M}_{\mathcal{T}}) + \frac{\kappa}{n} + (2\kappa + 1)\sqrt{\frac{\ln 1/\delta}{2n}},$$

where $\epsilon(\boldsymbol{M}_{\mathcal{T}})$ is the generalization loss and $\epsilon_{\mathcal{T}}(\boldsymbol{M}_{\mathcal{T}})$ its empirical estimate over $\mathcal{T}$.

Analysis For convenience, given a bilinear model $S_{\boldsymbol{M}}$, we denote by $\boldsymbol{M}_R$ both the similarity defined by the matrix $\boldsymbol{M}$ and its associated set of reasonable points R (when it is clear from the context we may omit the subscript R). Given a similarity $\boldsymbol{M}_R$, we use ℓ to define the loss over one example and we define the *error over the true distribution* by

$$\epsilon(\boldsymbol{M}_R) = \mathbb{E}_{\boldsymbol{z}=(\boldsymbol{x},y)\sim\mu}\ell(\boldsymbol{M}, \boldsymbol{z}, R).$$

The empirical error over the sample $\mathcal{T}$ is defined as

$$\epsilon_{\mathcal{T}}(\boldsymbol{M}_R) = \frac{1}{n}\sum_{i=1}^{n} \ell(\boldsymbol{M}, \boldsymbol{z}_i, R).$$

In our context, to prove the uniform stability property we need to show:

$$\forall \mathcal{T}, \forall i, \sup_{z} |\ell(\boldsymbol{M}, z, R) - \ell(\boldsymbol{M}^i, z, R^i)| \leq \frac{\kappa}{n},$$

where $\boldsymbol{M}$ is learned from $\mathcal{T}$, $R \subseteq \mathcal{T}$, $\boldsymbol{M}^i$ is the matrix learned from $\mathcal{T}^i$ and $R^i \subseteq \mathcal{T}^i$ is the set of reasonable points associated to $\mathcal{T}^i$. Note that R and R^i are of equal size and can differ in at most one example, depending on whether z_i or z'_i belong to their corresponding set of reasonable points and note that ℓ is bounded by 1. To be able to show the above property, we need the following results.

Lemma 8.31 *For any labeled examples* $z = (\boldsymbol{x}, y)$, $z' = (\boldsymbol{x}', y')$ *and any models* $\boldsymbol{M}_R$, $\boldsymbol{M}'_{R'}$, *the following properties hold:*

P1: $|S_{\boldsymbol{M}}(\boldsymbol{x}, \boldsymbol{x}')| \leq 1$,

P2: $|S_{\boldsymbol{M}}(\boldsymbol{x}, \boldsymbol{x}') - S_{\boldsymbol{M}'}(\boldsymbol{x}, \boldsymbol{x}')| \leq \|\boldsymbol{M} - \boldsymbol{M}'\|_{\mathcal{F}}$,

P3: $|\ell(\boldsymbol{M}, z, R) - \ell(\boldsymbol{M}', z, R')| \leq 1|\frac{\sum_{k=1}^{n_R} y_k S_{\boldsymbol{M}}(\boldsymbol{x}, \boldsymbol{x}_k)}{\gamma n_R} - \frac{\sum_{j=1}^{n_{R'}} y'_k S_{\boldsymbol{M}'}(\boldsymbol{x}, \boldsymbol{x}'_k)}{\gamma n_{R'}}|$ *(1-admissibility property of* ℓ*).*

The proof is given in Appendix A.3.1. Let $F_{\mathcal{T}} = \frac{1}{n}\sum_{i=1}^{n} \ell(\boldsymbol{M}, z_i, R) + \lambda\|\boldsymbol{M}\|_{\mathcal{F}}^2$ be the objective function of (8.10) with respect to a sample $\mathcal{T}$ and a set of reasonable points $R \subseteq \mathcal{T}$. The following lemma bounds the deviation between $\boldsymbol{M}$ and $\boldsymbol{M}^i$.

Lemma 8.32 *For any* $\boldsymbol{M}$ *and* $\boldsymbol{M}^i$ *that are minimizers of* $F_{\mathcal{T}}$ *and* $F_{\mathcal{T}^i}$ *respectively, we have:*

$$\|\boldsymbol{M} - \boldsymbol{M}^i\|_{\mathcal{F}} \leq \frac{1}{\lambda n \gamma}.$$

The proof is given in Appendix A.3.2.

Applying Theorem 8.30 with Lemma 8.32 leads to the main result.

Theorem 8.33 *Let* $\gamma > 0$, $\delta > 0$ *and* $n > 1$. *With probability at least* $1 - \delta$, *for any model* $\boldsymbol{M}_R$ *learned with SLLC, we have:*

$$\epsilon(\boldsymbol{M}_R) \leq \epsilon_{\mathcal{T}}(\boldsymbol{M}_R) + \frac{1}{n}\left(\frac{\hat{\tau} + 2\lambda\gamma}{\hat{\tau}\lambda\gamma^2}\right) + \left(\frac{2(\hat{\tau} + 2\lambda\gamma)}{\hat{\tau}\lambda\gamma^2} + 1\right)\sqrt{\frac{\ln 1/\delta}{2n}}.$$

Theorem 8.33 is both a consistency bound for the learned similarity and a generalization guarantee for the linear classifier using it. Indeed, by minimizing $\epsilon_{\mathcal{T}}$, one minimizes ϵ and thus an upper bound on the true risk of the resulting linear classifier, as stated by Theorem 8.28.

8.3.2 BOUNDS BASED ON RADEMACHER COMPLEXITY

Another strategy to derive generalization bounds is to use analysis based on Rademacher complexity [Bartlett and Mendelson, 2002]. In statistical learning theory, the Rademacher complexity measures the expressiveness of a hypothesis set composed of real-valued functions. Cao et al. [2012a] have extended this notion to metric and similarity learning. This framework is very expressive and allows one to prove generalization bounds for a large class of metric and similarity learning algorithms with various matrix norm regularizers.

To illustrate this approach, we present briefly the work of Guo and Ying [2014] in the context of bilinear similarity learning where the guarantees are obtained in two steps. First a generalization bound over the learned similarity itself is derived. Then, this result is used to obtain generalization guarantees for a linear classifier based on the learned similarity.

Preliminaries

As in the previous section, we consider the bilinear similarity S_M but the $d \times d$ real matrix $\boldsymbol{M}$ is constrained to be symmetric PSD.[3] Let $\mathcal{T} = \{z_i = (\boldsymbol{x}_i, y_i)\}_{i=1}^n$ be a training sample. We consider the following similarity learning problem:

$$\begin{aligned} \min_{\boldsymbol{M} \in \mathbb{S}_+^d} \quad & \frac{1}{n}\sum_{i=1}^{n} \ell(\boldsymbol{M}, z_i, \mathcal{T}) \quad + \quad \lambda \|\boldsymbol{M}\|_{\mathcal{F}}^2 \\ \text{s.t.} \quad & \ell(\boldsymbol{M}, z_i, \mathcal{T}) = [1 - y_i \frac{1}{\gamma n} \sum_{j=1}^{n} y_j S_{\boldsymbol{M}}(\boldsymbol{x}_i, \boldsymbol{x}_j)]_+, \end{aligned} \tag{8.11}$$

where $\gamma > 0$ is the margin. The slight difference with formulation (8.10) is that the set of reasonable points is taken to be the entire training sample $\mathcal{T}$.

As usual, the *generalization error* is of a bilinear similarity represented by a matrix $\boldsymbol{M}$ is defined with respect to a distribution μ over $\mathcal{Z} = \mathcal{X} \times \mathcal{Y}$ such that:

$$\epsilon(\boldsymbol{M}) = \mathbb{E}_{(\boldsymbol{x},y)\sim\mu}[1 - \frac{1}{\gamma}\mathbb{E}_{(\boldsymbol{x}',y')\sim\mu} y y' S_{\boldsymbol{M}}(\boldsymbol{x}, \boldsymbol{x}')]_+.$$

The *empirical error* of a bilinear similarity represented by a matrix $\boldsymbol{M}$ over a sample $\mathcal{T}$ is defined as follows:

$$\epsilon_{\mathcal{T}}(\boldsymbol{M}) = \frac{1}{n}\sum_{i=1}^{n} \ell(\boldsymbol{M}, z_i, \mathcal{T}).$$

Before presenting the Rademacher generalization bound associated to a bilinear similarity learned by Formulation (8.11), we need some notations. For any matrix norm $\|\cdot\|$, we define

[3]This is because Guo and Ying [2014] want $S_{\boldsymbol{M}}$ to be a valid kernel so that it can be used in SVM.

its dual norm $\|\cdot\|^*$ such that for any matrix $\boldsymbol{B}$: $\|\boldsymbol{B}\|^* = \sup_{\|\boldsymbol{A}\|\le 1} \operatorname{tr}(\boldsymbol{B}^T \boldsymbol{A})$. We denote $X^* = \sup_{\boldsymbol{x},\boldsymbol{x}'\in\mathcal{X}} \|\boldsymbol{x}'\boldsymbol{x}^T\|^*$. We define the Rademacher average complexity $\mathcal{R}_n$ with respect to the dual matrix norm as follows.

Definition 8.34 Let $\mathcal{T} = \{z_i = (\boldsymbol{x}_i, y_i)\}_{i=1}^n$ be a learning sample of size n *i.i.d.* from μ, and let $\boldsymbol{\sigma} = \{\sigma_i\}_{i=1}^n \in \{-1, 1\}^n$ be n independent Rademacher random variables such that $P(\sigma_i = 1) = P(\sigma_i = -1) = \frac{1}{2}$, the Rademacher average $\mathcal{R}_n$ is defined as

$$\mathcal{R}_n = \mathbb{E}_{\mathcal{T}\sim\mu}\mathbb{E}_{\boldsymbol{\sigma}}\left[\sup_{\boldsymbol{x}\in\mathcal{X}}\left\|\frac{1}{n}\sum_{(\boldsymbol{x}_i,y_i)\in\mathcal{T}}\sigma_i y_i \boldsymbol{x}_i \boldsymbol{x}^T\right\|^*\right].$$

Generalization Guarantees for Similarity Learning

From Definition 8.34, we can now present the generalization bound based on Rademacher complexity.

Theorem 8.35 Generalization bound with Rademacher complexity. *Let $\delta > 0$ and $n > 1$. For any similarity $\boldsymbol{M}_{\mathcal{T}}$ learned by Formulation (8.11) from a learning sample $\mathcal{T}$ of size n i.i.d. from μ, with probability at least $1-\delta$:*

$$\epsilon(\boldsymbol{M}_{\mathcal{T}}) < \epsilon_{\mathcal{T}}(\boldsymbol{M}_{\mathcal{T}}) + \frac{6\mathcal{R}_m}{\gamma\lambda} + \frac{2X^*}{\gamma\lambda}\sqrt{\frac{2\ln 1/\delta}{n}},$$

where $\epsilon(\boldsymbol{M}_{\mathcal{T}})$ is the generalization loss and $\epsilon_{\mathcal{T}}(\boldsymbol{M}_{\mathcal{T}})$ its empirical estimate over $\mathcal{T}$.

The proof of this result can be found in [Guo and Ying, 2014]. We give now some applications of this theorem by providing some estimation of the Rademacher average $\mathcal{R}_n$ for different regularizers frequently used in similarity learning as presented in [Guo and Ying, 2014].

Example 8.36 Frobenius norm We consider the case where the regularizer corresponds to the Frobenius norm $\|\boldsymbol{M}\| = \|\boldsymbol{M}\|_{\mathcal{F}} = \sqrt{\sum_{i=1}^d \sum_{j=1}^d m_{ij}^2}$. The dual norm of the Frobenius norm is itself. As a consequence, $X^* = \sup_{\boldsymbol{x},\boldsymbol{x}'\in\mathcal{X}} \|\boldsymbol{x}'\boldsymbol{x}^T\|_F = \sup_{\boldsymbol{x}\in\mathcal{X}} \|\boldsymbol{x}\|_F^2$. The Rademacher average can then be rewritten as:

$$\mathcal{R}_n = \mathbb{E}_{\mathcal{T}\sim\mu}\mathbb{E}_{\boldsymbol{\sigma}}\left[\sup_{\boldsymbol{x}\in\mathcal{X}}\|\frac{1}{n}\sum_{(\boldsymbol{x}_i,y_i)\in\mathcal{T}}\sigma_i y_i \boldsymbol{x}_i \boldsymbol{x}^T\|_F\right]$$

By Cauchy's inequality, we have:

$$\begin{aligned}
\mathcal{R}_n &= \mathbb{E}_{\mathcal{T}\sim\mu}\mathbb{E}_{\sigma}\left[\sup_{\boldsymbol{x}\in\mathcal{X}}\|\boldsymbol{x}\|_F\|\frac{1}{n}\sum_{(\boldsymbol{x}_i,y_i)\in\mathcal{T}}\sigma_i y_i \boldsymbol{x}_i\|_F\right] \\
&\leq \sup_{\boldsymbol{x}\in\mathcal{X}}\|\boldsymbol{x}\|_F\ \mathbb{E}_{\mathcal{T}\sim\mu}\left[\sqrt{\mathbb{E}_{\sigma}\|\frac{1}{n}\sum_{(\boldsymbol{x}_i,y_i)\in\mathcal{T}}\sigma_i y_i \boldsymbol{x}_i\|_F^2}\right] \\
&= \sup_{\boldsymbol{x}\in\mathcal{X}}\|\boldsymbol{x}\|_F\ \mathbb{E}_{\mathcal{T}\sim\mu}\left[\sqrt{\sum_{(\boldsymbol{x}_i,y_i)\in\mathcal{T}}\|\boldsymbol{x}_i\|_F^2}\frac{1}{n}\right] \\
&\leq \sup_{\boldsymbol{x}\in\mathcal{X}}\|\boldsymbol{x}\|_F^2\frac{1}{\sqrt{n}}.
\end{aligned}$$

Note that it can be shown that the trace norm regularization leads to the same quantity $\mathcal{R}_n$ as the Frobenius norm [Guo and Ying, 2014].

Example 8.37 *L_1* **norm** We take the formulation where $\|\boldsymbol{M}\| = \|\boldsymbol{M}\|_1 = \sum_{i=1}^d\sum_{j=1}^d |m_{ij}|$. The dual norm of L_1 norm is the L_∞ norm and thus

$$X^* = \sup_{\boldsymbol{x},\boldsymbol{x}'\in\mathcal{X}}\sup_{1\leq l,k\leq d}|\boldsymbol{x}^l\boldsymbol{x}'^k| = \sup_{\boldsymbol{x}\in\mathcal{X}}\|\boldsymbol{x}\|_\infty^2$$

where $\boldsymbol{x}^l$ denotes the value of the l-th attribute of instance $\boldsymbol{x}$. Then, it can be shown [Guo and Ying, 2014] that

$$\mathcal{R}_n \leq 2\sup_{\boldsymbol{x}\in\mathcal{X}}\|\boldsymbol{x}\|_\infty^2\sqrt{\frac{e\log(d+1)}{n}}.$$

Example 8.38 *$L_{2,1}$* **norm** We consider the formulation where $\|\boldsymbol{M}\| = \|\boldsymbol{M}\|_{2,1} = \sum_{i=1}^d \|m^i\|_2$. The dual norm of the $L_{2,1}$ norm is the $L_{2,\infty}$ norm which implies that $X^* = \sup_{\boldsymbol{x},\boldsymbol{x}'\in\mathcal{X}}\|\boldsymbol{x}'\boldsymbol{x}^T\|_{(2,\infty)} = \sup_{\boldsymbol{x}\in\mathcal{X}}\|\boldsymbol{x}\|_F\sup_{\boldsymbol{x}'\in\mathcal{X}}\|\boldsymbol{x}'\|_\infty$. Then, it can be shown [Guo and Ying, 2014] that

$$\mathcal{R}_n \leq 2\sup_{\boldsymbol{x}\in\mathcal{X}}\|\boldsymbol{x}\|_F\ \sup_{\boldsymbol{x}'\in\mathcal{X}}\|\boldsymbol{x}'\|_\infty\sqrt{\frac{e\log(d+1)}{n}}.$$

Generalization bound for linear SVM

The generalization bound given in Theorem 8.35 provides an upper bound on the generalization error of a linear classifier produced by a linear Support Vector Machine (SVM) [Vapnik, 1998] defined as follows:

$$\min_{\alpha} \frac{1}{n} \sum_{(x_i, y_i) \in \mathcal{T}} \left[1 - y_i \sum_{(x_j, y_j) \in \mathcal{T}}^{n} \alpha_j y_j S_M(x_i, x_j) \right]_+ \quad \text{s.t. } \|\alpha\|_1 \leq 1/\gamma. \tag{8.12}$$

Let $f_M(\cdot)$ be the linear classifier defined by $f_M(\cdot) = \sum_{(x_j,y_j)\in\mathcal{T}}^{n} \alpha_j y_j S_M(\cdot, x_j)$ and let $\epsilon(f_M) = \mathbb{E}_{(x,y)\sim\mu}[1 - y f_M(x)]_+$ be the true generalization error associated with f_M. The relationship between the generalization bound on the learned similarity $S_{M_\mathcal{T}}$ and the classifier $f_{M_\mathcal{T}}$ making use of this similarity is given by the following theorem.

Theorem 8.39 *Let $S_{M_\mathcal{T}}$ and $f_{M_\mathcal{T}}$ be the similarity and the linear classifiers learned from Problems* (8.11) *and* (8.12) *respectively. Then, for any $0 < \delta < 1$, with probability at least $1 - \delta$, we have:*

$$\epsilon(f_{M_\mathcal{T}}) \leq \epsilon_\mathcal{T}(M_\mathcal{T}) + \frac{4\mathcal{R}_n}{\lambda\gamma} + \frac{2X^*}{\lambda\gamma}\sqrt{\frac{2\log(1/\delta)}{n}}.$$

This bound is difficult to compare with the bound proved in the previous section with uniform stability. Results obtained with uniform stability depend on a specific learning algorithm, while the bounds based on Rademacher analysis above hold uniformly for the considered hypothesis class.

CHAPTER 9

Applications

Metric learning can potentially be beneficial whenever the notion of metric or similarity between instances plays an important role. Consequently, recent work in many domains have taken advantage of the algorithmic and theoretical advances in metric learning to cover a large spectrum of applications, like image retrieval, face recognition, image annotation, music recommendation, cartoon synthesis, webpage archiving, text retrieval, to name a few. This chapter aims to provide a brief overview of those main domains of applications of metric learning. As we will see, some of the proposed methods directly make use of algorithms we already introduced in the previous chapters of this book (e.g., LMNN, ITML, etc.). Those methods typically differ by the choice of the constraints the metric has to satisfy, or by slight changes in the objective function (loss function and/or regularization). On the other hand, other methods design a specific metric learning algorithm dedicated to deal with the application at hand. For those approaches, we will provide the necessary material to allow the reader to get the big picture of the considered algorithms, and refer him/her to the specific papers for further details.

Note that this chapter does not present an exhaustive list of applications. We focus our attention on three large fields where metric learning has been shown to be the most useful: *Computer Vision, Information Retrieval* and *Bioinformatics*. The first two typically deal with numerical data. Because it is out of the scope of this book, we will not detail the preprocessing needed to encode data involved in the applications of these fields and simply assume that images, texts or videos come in the form of feature vectors. On the other hand, Bioinformatics manipulate data that are often represented in a structured form, like sequences, trees, graphs or time series. In this field, metric learning algorithms are typically based on structural distances such as the edit distance or the Dynamic Time Warping distance.

9.1 COMPUTER VISION

Metric learning has indisputably received a lot of attention from the computer vision community during the past few years, as evidenced by tutorials and workshops organized at ECCV 2010, ICCV 2011 or ACCV 2014. This trend can be justified by the fact that metrics play a key role at different steps of the computer vision process. Indeed, there is a great need of appropriate metrics not only to compare images or videos in ad-hoc representations—such as bags-of-visual-words [Li and Perona, 2005]—but also in the pre-processing step consisting in building these representations (for instance, visual words are usually obtained by means of clustering). For this reason, there exists a large body of metric learning literature dealing specifically with computer

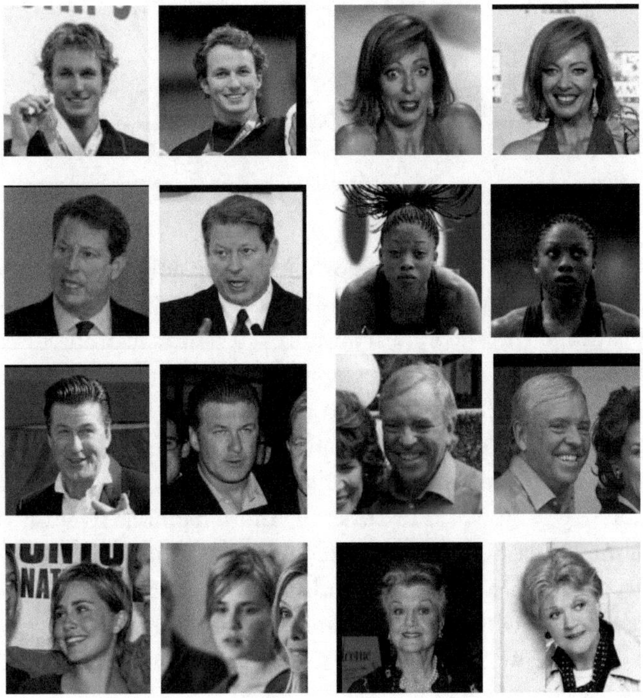

Figure 9.1: Some face images of the *Labeled Faces in the Wild* database which contains 13,233 labeled faces of 5,749 people taken from Yahoo!News. For a subset of 1,680 people, we have access to two or more images that allows the construction of positive pairs.

vision problems, such as image classification, object recognition, face recognition, kinship verification, visual tracking or image annotation. We present a brief overview of these methods in the following.

Face analysis Face analysis is a wide research field that covers many real-world problems, like face identification, sentiment recognition, gender classification, kinship verification, etc.

The problem of **face identification** is the visual identification task that aims to determine whether two face images depict the same person. Figure 9.1 shows some examples of faces drawn from the well-known *Labeled Faces in the Wild* dataset[1] [Huang et al., 2007] which contains (positive and negative) pairs of faces as well as individual labeled examples. We can notice that images describe a large range of possible variations in appearance (differences in pose, expression, background, etc.). Moreover, face identification models must be able to generalize to a new person who did not appear in the training set. This makes face identification a very difficult recognition

[1] `http://vis-www.cs.umass.edu/lfw.`

problem that received a lot of attention during the past few years, in particular in the context of metric learning.

Guillaumin et al. [2009b] present two Mahalanobis-like metric learning algorithms that allow a significant improvement on the *Labeled Faces in the Wild* dataset compared to the state of the art. The first one, called LDML, learns a linear transformation by optimizing the parameters of the following linear logistic discriminant model, which can be used to estimate whether two images depict the same person or not:

$$p(y_i = y_j|\boldsymbol{x}_i, \boldsymbol{x}_j, \boldsymbol{M}, b) = (1 + \exp(d_M(\boldsymbol{x}_i, \boldsymbol{x}_j) - b))^{-1},$$

where d_M is the Mahalanobis distance between two face images represented by feature vectors x_i and x_j, and b is a biased term. Both $\boldsymbol{M}$ and b are learned using a gradient ascent on the log-likelihood of $p(y_i = y_j|\boldsymbol{x}_i, \boldsymbol{x}_j, \boldsymbol{M}, b)$.

In [Cinbis et al., 2011], the authors use LDML to deal with the problem of face recognition in TV video. Their objective is to learn metrics that are robust to intra-person appearance variations. They apply LDML on tracks from episodes of the TV series "Buffy the Vampire Slayer" for which the authors annotated 639 face tracks.

It is worth noting that LDML requires labeled pairs of face images. Guillaumin et al. [2010] extend LDML to a weakly supervised setting, where the supervision occurs only at a *bag* level, i.e., where bags of face images are labeled with bags of labels. If two bags share at least one label, they are considered as being a positive pair.

In order to deal with non-linear face identification problems, Guillaumin et al. [2009b] present a second method, namely MkNN, based on a k-nearest neighbor (kNN) classifier. In its standard form, a kNN ckassifier approximates the Bayesian rule in a non-parameric setting by assigning to any feature vector x_i the most probable class c:

$$\arg\max_c p(y_i = c|\boldsymbol{x}_i) = \arg\max_k \frac{n_c^i}{k},$$

where n_c^i corresponds to the number of examples of class c in the k-neighborhood of x_i. To deal with pairs of examples, the authors extend $p(y_i = c|x_i)$ to the marginal probability $p(y_i = y_j|\boldsymbol{x}_i, \boldsymbol{x}_j)$, defined as follows:

$$p(y_i = y_j|\boldsymbol{x}_i, \boldsymbol{x}_j) = \frac{\sum_c n_c^i n_c^j}{k^2}.$$

Once again, this probability can be used to estimate whether two face images $\boldsymbol{x}_i$ and $\boldsymbol{x}_j$ depict the same person or not. To improve the performance of their model, the authors make use of LMNN (see Section 4.1.1) to determine the neighbors.

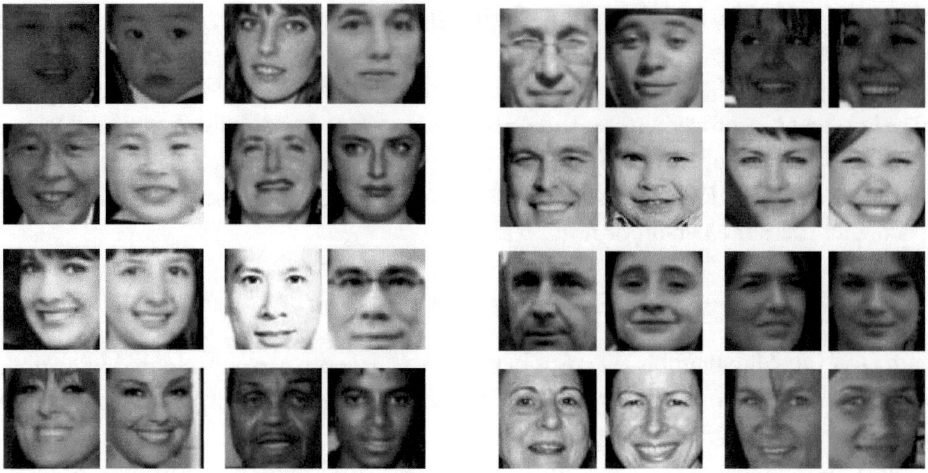

Figure 9.2: Some face images drawn from the KFW-I database (on the left) and KWI-2 database (ont the right). KFW-I and KFW-II differ by the fact that images of the former are collected from different photos while face images of KFW-II are collected from the same photo.

An even more challenging problem of face analysis is **kinship verification** from facial images (see [Fang et al., 2010]) for which very few databases are available. To address this issue, Lu et al. [2012a] created two new face datasets dedicated to this task, called KFW-I and KFW-II (see some examples in Figure 9.2). Moreover, inspired from LMNN, they proposed a neighborhood repulsed metric learning (NRML) method which moves closer pairs of kinship images and pushes away all the other examples in the neighborhood of the images of the pairs. More formally, given a set of N pairs $\{(\boldsymbol{x}_i, \boldsymbol{z}_i)\}_{i=1}^{N}$ of kinship images, the optimization problem takes the following form:

$$\max_{M} \quad \frac{1}{Nk}\sum_{i=1}^{N}\sum_{j=1}^{k} d_M^2(\boldsymbol{x}_i, \boldsymbol{z}_{ij}) + \frac{1}{Nk}\sum_{i=1}^{N}\sum_{j=1}^{k} d_M^2(\boldsymbol{x}_{ij}, \boldsymbol{z}_i) \quad \frac{1}{N}\sum_{i=1}^{N} d^2(\boldsymbol{x}_i, \boldsymbol{z}_i),$$

where $\boldsymbol{z}_{ij}$ (resp. $\boldsymbol{x}_{ij}$) corresponds to the jth k-nearest neighbor of $\boldsymbol{z}_i$ (resp. $\boldsymbol{x}_i$), and $d_M^2(\boldsymbol{x}_i, \boldsymbol{z}_i)$ is the Malahanobis distance between $\boldsymbol{x}_i$ and $\boldsymbol{z}_i$.

Note that the authors also propose a multiview NRML method of kinship verification to learn a common distance metric for measuring multiple feature representations of facial images.

Image classification Image classification is a very challenging task in computer vision that consists, for example, in identifying objects appearing in images. Many datasets are used by the community to compare new proposed methods such as PASCAL-VOC,[2] ImageNet [Deng

[2] http://pascallin.ecs.soton.ac.uk/challenges/VOC/.

et al., 2009] (see query images in Figure 9.3),[3] Caltech-256,[4] to cite a few. Metric learning has been shown to be very beneficial in image classification where distances between objects play an important role in the classification models.

In Wang et al. [2010], the authors present a metric learning algorithm that improves Image-to-Class (I2C) distance. Unlike Image-to-Image (I2I) distance, I2C distance computes the Euclidean distance between an image and a candidate class c. It typically compares keypoints of an image request X_i to nearest keypoints from class c, as follows:

$$d(\boldsymbol{X}_i, c) = \sum_{j=1}^{m} ||\boldsymbol{X}_{ij} - NN(\boldsymbol{X}_{ij}^c)||^2, \tag{9.1}$$

where $NN(\boldsymbol{X}_{ij}^c)$ is the nearest keypoints from class c of the jth keypoint $\boldsymbol{X}_{ij}$ of $\boldsymbol{X}_i$. The authors suggest learning a Mahalanobis distance $\boldsymbol{M}_c$ per class c and compute the learned I2C distance as follows:

$$d_{\boldsymbol{M}_c}(\boldsymbol{X}_i, c) = \sum_{j=1}^{m} (\boldsymbol{X}_{ij} - NN(\boldsymbol{X}_{ij}^c))^T \boldsymbol{M}_c (\boldsymbol{X}_{ij} - NN(\boldsymbol{X}_{ij}^c)), \tag{9.2}$$

Eq. (9.2) can be rewritten using matrix operations:

$$d_{\boldsymbol{M}}(\boldsymbol{X}_i, c) = \text{tr}(\Delta \boldsymbol{X}_{ic} \boldsymbol{M}_c \Delta \boldsymbol{X}_{ic}^T).$$

Then, using a direct adaptation of LMNN to the context of I2C, matrices $\boldsymbol{M}_1, ..., \boldsymbol{M}_c, ..., \boldsymbol{M}_C$ are learned by solving the following SDP optimization problem:

$$\begin{aligned} \min_{\boldsymbol{M}_1,...,\boldsymbol{M}_c,...,\boldsymbol{M}_C} \quad & (1-\lambda) \sum_{i,p\to i} \text{tr}(\Delta \boldsymbol{X}_{ip} \boldsymbol{M}_p \Delta \boldsymbol{X}_{ip}^T) + \lambda \sum_{i,p\to i,n\to i} \xi_{ipn} \\ \text{s.t.} \quad & \forall i,p,n : \text{tr}(\Delta \boldsymbol{X}_{in} \boldsymbol{M}_n \Delta \boldsymbol{X}_{in}^T) - \text{tr}(\Delta \boldsymbol{X}_{ip} \boldsymbol{M}_p \Delta \boldsymbol{X}_{ip}^T) \geq 1 - \xi_{ipn} \\ & \forall i,p,n : \xi_{ipn} \geq 0 \\ & \forall c : \boldsymbol{M}_c \succeq 0. \end{aligned} \tag{9.3}$$

The first constraint in (9.3) tends to ensure that the I2C distance from image $\boldsymbol{X}_i$ to its true class p is smaller than the distance to any other class n with a margin 1.

Experimental results on datasets Scene-15,[5] Sports,[6] Corel[7] and Caltech 101[8] show significant improvements when compared to the standard I2C distance in a Naive-Bayes Nearest-Neighbor classifier [Boiman et al., 2008].

[3]http://www.image-net.org
[4]http://www.vision.caltech.edu/Image_Datasets/Caltech256/
[5]http://www-cvr.ai.uiuc.edu/ponce_grp/data/.
[6]http://vision.stanford.edu/lijiali/event_dataset/
[7]http://vision.stanford.edu/resources_links.html
[8]http://www.vision.caltech.edu/Image_Datasets/Caltech101/

Figure 9.3: Examples of ImageNet query images.

Image Annotation Given an image, image annotation aims at automatically assigning a set of relevant labels or keywords that can be used, for example, as tags. Mensink et al. [2012] deal with large-scale annotation by treating it as a multi-label classification problem. Two new metric learning-based methods are presented to deal with this task. The first one aims to improve LMNN by overcoming the problem of having to a priori select target neighbors using the L_2-norm in the original representation space. To do so, they slightly modify LMNN and suggest making use of the current learned metric to select those target instances in the projected space. The second is based on the Nearest-Class Mean (NCM) classifier where mean feature vectors are used to represent their corresponding classes. The authors introduce a new metric learning algorithm based on multi-class logistic discrimination which constrains images to be closer to their class mean $_c$ than to any other class mean. A stochastic gradient descent algorithm is used to optimize the parameters using only a small subset of images at each step. More formally, let $p(c|\boldsymbol{x})$ be the probability that an example $\boldsymbol{x}$ belongs to class c:

$$p(c|\boldsymbol{x}) = \frac{\exp(\ d^2_M(\ _c, \boldsymbol{x}))}{\sum_{c'=1}^{C} \exp(\ d^2_M(\ _c, \boldsymbol{x}))}.$$

Minimizing the negative log-likelihood of $p(c|\boldsymbol{x})$, we get the following gradient

$$\frac{2}{N} \sum_{i=1}^{N} \sum_{c=1}^{C} ([\![y_i = c]\!] \quad p(c|\boldsymbol{x}))\, \boldsymbol{M}(\ _c \quad \boldsymbol{x}_i)(\ _c \quad \boldsymbol{x}_i)^T.$$

The authors report good results on the ImageNet Large Scale Visual Recognition Challenge 2010 dataset[9] composed of 1.2M training images belonging to 1,000 classes.

In Guillaumin et al. [2009a], tags are automatically predicted using a weighted nearest-neighbor model, called TagProp. The weights of the neighbors are optimized via a Mahalanobis distance learning method which maximizes the likelihood of the tags over the training set. Let $y_{iw} \in \{\ 1, +1\}$ describe the presence or absence of keyword w in image $\boldsymbol{x}_i$. The model for predicting the presence of a keyword in an image is defined as follows:

[9] `http://www.image-net.org/challenges/LSVRC/2010/`.

$$p(y_{iw} = +1) = \sum_j \pi_{ij} p(y_{iw} = +1|\boldsymbol{x}_j)$$

with

$$p(y_{iw} = +1|\boldsymbol{x}_j) = \begin{cases} 1-\epsilon \\ \epsilon \end{cases}$$

where ϵ is set to a small value and π_{ij} is the weight of image $\boldsymbol{x}_j$ for predicting the tags of image $\boldsymbol{x}_i$. To learn the weights π_{ij}, the authors suggest to define them in the following Mahalanobis distance-based form:

$$\pi_{ij} = \frac{\exp(-d_M(\boldsymbol{x}_i, \boldsymbol{x}_j))}{\sum_{j'} \exp(-d_M(\boldsymbol{x}_i, \boldsymbol{x}'_j))}.$$

Experiments on three challenging image datasets (Corel 5k, ESP Game, IAPR TC12) show significant improvements over uniform weighting.

Other methods Note that many other metric learning algorithms have been presented in the literature and tested on image datasets. Xiao and Madabhushi [2011] propose an aggregated distance metric learning method, inspired from the boosted BDM method [Yang et al., 2010] and from bagging, which deals with situations where the number of training images is limited. Experiments are performed on X-ray images from ImageCLEFmed 2009 dataset.[10] Lu et al. [2012b] deal with gait-based gender classification from videos. Since point-to-point distances are known to be irrelevant to characterize gait images, they suggest to optimize a point-to-set distance that minimizes the intraclass and maximizes the interclass variations. Experiments achieved from a dataset of gait sequences of 20 subjects captured by a Microsoft Kinect depth camera show significant improvement in comparison with ITML, LMNN and NCA. In Wang et al. [2012a], semantic relations of images are used to improve the metric learned by NCA. The authors use a semantic measure based on the WordNet database. Experiments on Caltech256 and ImageNet datasets show that the proposed method improves NCA as well as LMNN. In [Yang and Jin, 2006], a local distance metric algorithm (LDM) is introduced that is particularly well-suited when classes are described by multimodal data distributions. LDM moves closer pairs of points that belong to the same mode and pushes away pairs from different classes. Experiments on the Corel dataset shows the interest of learning such local metrics. Metric learning for visual category recognition is addressed in Frome et al. [2007] where local image-to-image distances, in the form of weighted sums of distances between patches, are learned and are shown to be globally consistent. Experiments are performed on the Caltech 101 dataset. Nowak and Jurie [2007] learn a similarity function between local image patches in order to compare never seen objects. Experiments are conducted on four datasets: Toy Cars,[11] Ferencz & Malik cars,[12] Jain

[10]http://www.imageclef.org/.
[11]http://lear.inrialpes.fr/people/nowak/.
[12]http://www.eecs.berkeley.edu/Research/Projects/CS/vision/shape/vid/dataset.html.

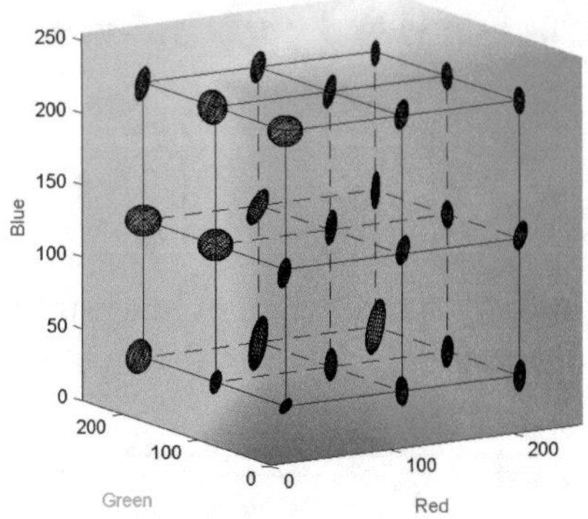

Figure 9.4: The RGB cube has been splitted into 20 regular regions, where the surface of each represented ellipsoid corresponds to the RGB colors lying at the corresponding learned local perceptual distance of 1 from the center of the ellipsoid.

Faces[13] and Coil-100.[14] Li et al. [2012] present an online algorithm to optimize a Mahalanobis distance used in a visual tracker. Finally, Perrot et al. [2014] propose to learn perceptual color distances. It is well known that the RGB cube is not perceptually uniform, that is, two equidistant pairs of pixels do not correspond to the same color difference perceived by human observers. In a linear regression setting, the authors learn local Mahalanobis distances that allow capturing not only some non linearity in the data but also local distortions of the RGB cube (see Figure 9.4).

9.2 BIOINFORMATICS

Many problems in bioinformatics involve comparing sequences such as DNA, protein or temporal series. These comparisons are based on structured metrics such as edit-based distance measures (or related to string alignment scores) for strings or Dynamic Time Warping distance for temporal series. Learning these metrics to adapt them to the task of interest can greatly improve the results. In this context, remote homology detection between protein sequences is a key problem in computational biology. There already exist some matrices that are dedicated to score amino acid matches. For example, the BLOSUM (BLOcks SUbstitution Matrix) matrices are generated from the Blocks database and are used for sequence alignment of proteins. Several BLOSUM matrices, named with numbers, have been constructed according to different alignment databases. Matrices with high numbers are devoted to compare closely related sequences, while matrices with small numbers are dedicated to compare distant related sequences (see, for exam-

[13] `http://vis-www.cs.umass.edu/fddb/`.
[14] `http://www.cs.columbia.edu/CAVE/software/softlib/coil-100.php`.

	Ala	Arg	Asn	Asp	Cys	Gln	Glu	Gly	His	Ile	Leu	Lys	Met	Phe	Pro	Ser	Thr	Trp	Tyr	Val
Ala	4																			
Arg	-1	5																		
Asn	-2	0	6																	
Asp	-2	-2	1	6																
Cys	0	-3	-3	-3	9															
Gln	-1	1	0	0	-3	5														
Glu	-1	0	0	2	-4	2	5													
Gly	0	-2	0	-1	-3	-2	-2	6												
His	-2	0	1	-1	-3	0	0	-2	8											
Ile	-1	-3	-3	-3	-1	-3	-3	-4	-3	4										
Leu	-1	-2	-3	-4	-1	-2	-3	-4	-3	2	4									
Lys	-1	2	0	-1	-3	1	1	-2	-1	-3	-2	5								
Met	-1	-1	-2	-3	-1	0	-2	-3	-2	1	2	-1	5							
Phe	-2	-3	-3	-3	-2	-3	-3	-3	-1	0	0	-3	0	6						
Pro	-1	-2	-2	-1	-3	-1	-1	-2	-2	-3	-3	-1	-2	-4	7					
Ser	1	-1	1	0	-1	0	0	0	-1	-2	-2	0	-1	-2	-1	4				
Thr	0	-1	0	-1	-1	-1	-1	-2	-2	-1	-1	-1	-1	-2	-1	1	5			
Trp	-3	-3	-4	-4	-2	-2	-3	-2	-2	-3	-2	-3	-1	1	-4	-3	-2	11		
Tyr	-2	-2	-2	-3	-2	-1	-2	-3	2	-1	-1	-2	-1	3	-3	-2	-2	2	7	
Val	0	-3	-3	-3	-1	-2	-2	-3	-3	3	1	-2	1	-1	-2	-2	0	-3	-1	4

Figure 9.5: BLOSUM62 matrix. Each row/column corresponds to an amino acid.

ple, Blosum62 matrix in Figure 9.5). On the other hand, the PAM (Point Accepted Mutation) matrices are based on the probability of single point mutations.

In Saigo et al. [2006], the authors use BLOSUM and PAM as starting matrices for optimizing a string alignment kernel specifically dedicated to detect remote homologies. As already mentioned in Chapter 7, this LA (Local Alignment) kernel is computed from a matrix of substitution between amino acids. It takes into account all the possible local alignments for changing X into X′, where an alignment score $s(\mathsf{X}, \mathsf{X}', \;)$ is used instead of the standard edit distance:

$$k_{LA}(\mathsf{X}, \mathsf{X}') = \sum_{\phi} e^{t\, s(\mathsf{X}, \mathsf{X}', \phi)},$$

where t is a parameter. The score $s(\mathsf{X}, \mathsf{X}', \;)$ of an alignment is defined as:

$$s(\mathsf{X}, \mathsf{X}', \;) = \sum_{u,v \in \chi} n_{u,v}(\mathsf{X}, \mathsf{X}', \;)\ C_{uv} \quad n_{g_d}(\mathsf{X}, \mathsf{X}', \;)\ g_d \quad n_{g_e}(\mathsf{X}, \mathsf{X}', \;)\ g_e,$$

where $n_{u,v}(\mathsf{X}, \mathsf{X}', \;)$ is the number of times that symbol u is aligned with v while g_d and g_e, along with their corresponding number of occurrences $n_{g_d}(\mathsf{X}, \mathsf{X}', \;)$ and $n_{g_e}(\mathsf{X}, \mathsf{X}', \;)$, are two parameters dealing respectively with the opening and extension of gaps. The edit costs C_{uv} as well as the gap parameters are learned by gradient descent. The optimized substitution matrix is then used to efficiently separate true homologs from non-homologs in a set of sequences issued from the COG database.

In Boyer et al. [2009], the authors deal with the detection of Transcription Factor Binding Sites (TFBS) in sequences of promoters of orthologous genes, i.e., genes having the same function and related by descent from a common ancestor. It has been demonstrated that TFBS are under evolutionary selection, meaning that they should have evolved much more slowly than other non-coding sequences. This difference in evolution speed can be observed by comparing sequences of orthologous genes between sufficiently distant species. In terms of edit distance, this means that such TFBS should be at a closer edit distance than other regions. The authors propose an approach that learns the parameters (the edit costs) of a Constrained State Machine, where the choice of a transition between two states of the automaton is driven by constraints fulfilled by the input string. This allows the model to take into account background knowledge during the learning step.

Definition 9.1 A *Constrained State Machine* (CSM) is a tuple $\langle \Sigma, Q, C, T, \delta, \pi \rangle$ where Σ is a finite alphabet, Q is a set of states, C is a set of constraints and:

- $T : Q \times C \times Q \rightarrow [0, 1]$ defines the probability of a transition. $T(q_j | q_i, c_{q_i,k}(X, t))$ will denote the probability of going to state q_j given that we are currently in state q_i and that the constraint $c_{q_i,k}(X, t)$ is satisfied. For a given state q_i and a constraint $c_{q_i,k}$, the outgoing transitions must fulfill the following condition:

$$\sum_{q_j \in Q} T(q_j | q_i, c_{q_i,k}) = 1. \tag{9.4}$$

- δ is a family of $|Q|$ matrices. Each element δ_{q_i} is the matrix of edit probabilities of state q_i.
- $\pi : Q \rightarrow [0, 1]$ is the initial probability function which must satisfy the following statistical condition:

$$\sum_{q \in Q} \pi(q) = 1. \tag{9.5}$$

The parameters of the CSM are optimized using an EM-based approach from a set of 13,520 pairs of orthologous promoter sequences.

Detecting analogous enzymes based on the local structures of active sites is also a hot research topic in Bioinformatics. Kato and Nagano [2010] proposed a metric learning algorithm to automatically detect catalytic sites. The goal is to search for proteins $\boldsymbol{X}'$ with a local structure composed of n atoms that is similar to a query template $\boldsymbol{X}$. The learned metric takes the form of the following weighted mean square deviation:

$$E(\boldsymbol{X}, \boldsymbol{X}', \boldsymbol{R}, \boldsymbol{v}, \boldsymbol{w}) = \sum_{i=1}^{n} w_i ||\boldsymbol{x}_i - (\boldsymbol{R}\boldsymbol{x}'_i + \boldsymbol{v})||^2,$$

where $\boldsymbol{R}$ is a rotation matrix, $\boldsymbol{v}$ a translation and $\boldsymbol{w}$ the weight vector to be optimized by the following problem:

$$\begin{aligned}
\min_{\theta,\xi,w} \quad & \frac{1}{\mathcal{S}}\sum_{i\in\mathcal{S}}\xi_i + \frac{1}{\mathcal{D}}\sum_{i\in\mathcal{D}}\xi_i \\
\text{s.t.} \quad & \forall i : y_i(E(\boldsymbol{X}, \boldsymbol{X}'^i, \boldsymbol{R}_i, \boldsymbol{v}_i, \boldsymbol{w}) - \theta) \leq \xi_i,
\end{aligned} \tag{9.6}$$

where ξ_i are nonnegative slack variables, and $\mathcal{S}$ (resp. $\mathcal{D}$) is a set of site matches (resp. dismatches). Problem (9.6) can be reduced to a linear program.

Note also that Wang et al. [2012b] propose a protein-to-protein dissimilarity learning method that takes into account the contextual information of a protein, i.e., its close-neighborhood. The notion of similarity relies on the hypothesis that two proteins having close contexts should have a high similarity score. Finally, Xiong and Chen [2006] introduce a k-nearest neighbors-based metric learning algorithm for cancer classification using microarray data which optimizes a data-dependent kernel.

9.3 INFORMATION RETRIEVAL

The objective of many information retrieval systems, such as search engines, is to provide the user with the most relevant documents (image, text) according to his/her query. This ranking is often achieved by using a metric between two documents or between a document and a query. Therefore, optimizing such metrics can be very beneficial. It is shown in Lee et al. [2008] that the information retrieval (IR) field requires specialized metric learning algorithms. Indeed, as seen so far, standard approaches assume that the distance between two similar objects has to be smaller than the distance between two dissimilar objects. It turns out that this assumption does not hold in IR especially when some queries are far away from all the elements of a given database while others are close to many of the elements of the same database. To overcome this problem, Lee et al. [2008] present a rank-based distance metric learning where a Mahalanobis-like metric $d(\boldsymbol{x}, \boldsymbol{x}', \boldsymbol{M})$ is optimized and regularized by making use of the Burg matrix divergence $D(\boldsymbol{M}, \hat{\boldsymbol{M}}_0)$ between $\boldsymbol{M}$ and a matrix $\boldsymbol{M}_0$ that may contain domain knowledge. This method only considers pairwise constraints issued from the same query. In other words, a must-link constraint ($y = +1$) has to lead to a smaller distance than a cannot-link constraint ($y = -1$) only if those two constraints come from the same query. The optimization problem takes the following form:

$$\begin{aligned}
\min_{\boldsymbol{M}\succeq 0,\xi\geq 0} \quad & \sum_{i=1}^{N_Q}\sum_{k,j=1}^{K} \delta(y_{ij}, -1)\delta(y_{ik}, +1)\xi^i_{jk} + \frac{\lambda}{2}D(\boldsymbol{M}, \hat{\boldsymbol{M}}_0) \\
\text{s.t.} \quad & d(\boldsymbol{q}_i, \boldsymbol{x}_{ij}, \boldsymbol{M}) - d(\boldsymbol{q}_i, \boldsymbol{x}_{ik}, \boldsymbol{M}) \geq 1 - \xi^k_{jk} \\
& \xi^i_{jk} \geq 0 \text{ and } \boldsymbol{M} \succeq 0,
\end{aligned}$$

where $\delta(y, a)$ is a Dirac function that gives 1 if $y = a$ and 0 otherwise, ϵ^i_{jk} are slack variables, N_Q is the number of queries used to build the pairwise constraints and K is a parameter which corresponds to the top K examples that are retrieved. Experiments are performed in tatoo image retrieval for suspect or victim identification.

Unlike many other applications, text analysis is a very high dimensional problem where d can reach the size of tens of thousands of features. As we have seen in Chapter 4, many Mahalanobis distance learning algorithms have an $O(d^3)$ complexity due to the PSD constraint, making intractable in this setting. Similarity learning without PSD constraint has thus attracted a lot of interest, often with similarity function of the form

$$S_{\boldsymbol{M}}(\boldsymbol{x}, \boldsymbol{x}') = \frac{\boldsymbol{x}^T \boldsymbol{M} \boldsymbol{x}'}{N(\boldsymbol{x}, \boldsymbol{x}')},$$

where $\boldsymbol{M} \in \mathbb{R}^{d \times d}$ is not required to be PSD nor symmetric, and $N(\boldsymbol{x}, \boldsymbol{x}')$ is a normalization term which depends on $\boldsymbol{x}$ and $\boldsymbol{x}'$. Note that $S_{\boldsymbol{M}}$ can be seen as a generalization of the cosine similarity, widely used in text and image retrieval [see for instance Baeza-Yates and Ribeiro-Neto, 1999, Sivic and Zisserman, 2009], where $\boldsymbol{M}$ is the identity matrix and $N(\boldsymbol{x}, \boldsymbol{x}') = ||\boldsymbol{x}||_2||\boldsymbol{x}'||_2$.

In Qamar et al. [2008], the authors learn $S_{\boldsymbol{M}}(\boldsymbol{x}, \boldsymbol{x}')$ by defining "target neighbors" and using an LMNN-like fashion. They optimize the similarity in an online manner with an algorithm based on voted perceptron. They apply their SiLA algorithm on the 20-newsgroups dataset, composed of about 20,000 posted articles from 20 newsgroups.

In Mikawa et al. [2011], the authors learn $S_{\boldsymbol{M}}(\boldsymbol{x}, \boldsymbol{x}')$ where $N(\boldsymbol{x}, \boldsymbol{x}') = ||\boldsymbol{x}||_2||\boldsymbol{x}'||_2$. Using $\boldsymbol{d}_1, \ldots, \boldsymbol{d}_D$ documents coming from $c_1, \ldots, c_N$ categories, they optimize the matrix $\boldsymbol{M}$ so as to maximize the similarities between documents and the centroid $\boldsymbol{g}_n$ of their corresponding categories. The optimization problem is defined as follows:

$$\begin{aligned} \max_{\boldsymbol{M}} \quad & \sum_{n=1}^{N} \sum_{\boldsymbol{d}_i \in c_n} S_{\boldsymbol{M}}(\boldsymbol{d}_i, \boldsymbol{g}_n) \\ \text{s.t.} \quad & \det(\boldsymbol{M}) = 1. \end{aligned}$$

In order to capture the specificities of each category $c_1, \ldots, c_N$ of documents, Mikawa et al. [2012] extend the above work by learning N individual matrices $\boldsymbol{M}_i$.

To further scale with the dimension, Davis and Dhillon [2008] propose a method that scales linearly with d by searching for a compressed representation of the data which is only composed of d parameters. This is achieved by a semi-supervised variant of latent semantic analysis (LSA) which projects the documents in a lower-dimensional space.

The idea of projecting the data in a lower space is also used by Lebanon [2006] in the context of classification of text documents. The author learns a Riemannian metric by seeking a lower dimensional representation of the documents. Assuming that the dimension of the submanifold depends on the location of the data, he learns a metric among a parametric family of metric candidates which can capture local variations. Experiments are performed on the WebKB dataset where documents are described by a term-frequency representation.

Finally, in order to take into account the semantic relatedness between terms, Yih et al. [2011] propose a Similarity Learning via Siamese Neural Network (S2Net) that learns a projection matrix that maps the term-vectors onto the underlying concept space of the text documents. This is efficiently done, even for high-dimensional problems, by training, w.r.t. the cosine measure, the parameters of a two layers-neural network. The neurons of the input layer encode the raw term vector while the output layer represents the learned lower-dimensional concept space. The authors use the logistic loss over relative constraints Δ of the following form:

$$\Delta = S_M(\boldsymbol{x}_i, \boldsymbol{x}_j) - S_M(\boldsymbol{x}_k, \boldsymbol{x}_l),$$

where term vectors $\boldsymbol{x}_i$ and $\boldsymbol{x}_j$ are supposed to be more similar than $\boldsymbol{x}_k$ and $\boldsymbol{x}_l$.

CHAPTER 10
Conclusion

In this book, we provided a comprehensive review of the main methods, theoretical frameworks and applications of metric learning. We briefly summarize the current state of the art, and attempt to draw promising lines for future research.

10.1 SUMMARY

Numerical data Metric learning for numerical data has reached a good level of maturity. This is illustrated in Table 10.1, which summarizes the properties of the main algorithms presented in this book. Indeed, there exist methods to learn various types of distance and similarity functions: linear and nonlinear, multiple local metrics, histogram distances, etc. Metric learning can also deal with a large spectrum of practical settings, such as multi-task and transfer learning, semi-supervised learning and domain adaptation. Furthermore, we have seen that recent efforts have been put into scaling-up metric learning to big data scenarios where the number of training samples and/or the dimensionality of the data can be very large. Finally, statistical learning theory for metric learning has received some attention, with several frameworks to derive generalization guarantees for the learned metric and, in some particular cases, for the classifier based on the metric.

Structured data On the other hand, much less work has gone into metric learning for structured data, as summarized in Table 10.2. We have presented some work on metric learning on strings, trees, graphs and time series, but much of the recent progress in metric learning for numerical data has not yet propagated to the structured case. Indeed, most approaches remain based on EM-like algorithms which make them intractable for large-scale data and hard to analyze theoretically due to local optima. Settings of practical importance like multi-task learning and domain adaptation have not been addressed. Nevertheless, recent work such as GESL [Bellet et al., 2011] and MLTSA [Lajugie et al., 2014b] have shown that drawing inspiration from successful feature vector formulations can be beneficial in terms of scalability and flexibility, even though it may require a simplification of the structured metric. This is a promising direction and probably a good omen for the development of this research topic.

10.2 OUTLOOK

In light of this book, we can identify the limitations of the current literature and speculate on where the future of metric learning is going.

Table 10.1: Main features of metric learning methods for feature vectors. Scalability levels are relative and given as a rough guide.

Page	Name	Year	Source Code	Supervision	Form of Metric	Scalability w.r.t. n	Scalability w.r.t. d	Optimum	Dimension Reduction	Regularizer	Additional Information
18	MMC	2002	Yes	Weak	Linear	★☆☆	☆☆☆	Global	No	None	—
18	NCA	2004	Yes	Full	Linear	★☆☆	★★☆	Local	Yes	None	For k-NN
19	LMNN	2005	Yes	Full	Linear	★★☆	☆☆☆	Global	No	None	For k-NN
21	S&J	2003	No	Weak	Linear	★★☆	★★★	Global	No	Frobenius norm	—
22	ITML	2007	Yes	Weak	Linear	★☆☆	★☆☆	Global	No	LogDet	Online version
23	SML	2009	No	Weak	Linear	★☆☆	☆☆☆	Global	Yes	$L_{2,1}$ norm	—
23	BoostMetric	2009	Yes	Weak	Linear	★☆☆	★☆☆	Global	Yes	None	—
27	POLA	2004	No	Weak	Linear	★★★	★☆☆	Global	No	None	Online
27	RDML	2009	No	Weak	Linear	★★★	★★☆	Global	No	Frobenius norm	Online
28	OASIS	2009	Yes	Weak	Linear	★★★	★★☆	Global	No	Frobenius norm	Online
34	LSMD	2005	No	Weak	Nonlinear	★★☆	★★☆	Local	Yes	None	—
35	GB-LMNN	2012	Yes	Full	Nonlinear	★★☆	★★☆	Local	Yes	None	—
36	HDML	2012	Yes	Weak	Nonlinear	★★☆	★★☆	Local	Yes	L_2 norm	Hamming distance
37	M^2-LMNN	2008	Yes	Full	Local	★★☆	☆☆☆	Global	No	None	—
37	PLML	2012	Yes	Weak	Local	★★☆	☆☆☆	Global	No	Manifold+Frob	—
39	SCML	2014	Yes	Weak	Local	★★☆	★★☆	Local	Yes	$L_{2,1}$ norm	+global/multi-task
40	Bk-means	2009	No	Weak	Local	★☆☆	★★★	Global	No	RKHS norm	Bregman dist.
41	RFD	2012	Yes	Weak	Local	★★☆	★★★	N/A	No	None	Random forests
42	mt-LMNN	2010	Yes	Full	Linear	★★☆	☆☆☆	Global	No	Frobenius norm	Multi-task
43	MLCS	2011	No	Weak	Linear	N/A	★★☆	Local	Yes	N/A	Multi-task
43	GPML	2012	No	Weak	Linear	N/A	★★☆	Global	Yes	von Neumann	Multi-task
44	TML	2010	Yes	Weak	Linear	★★☆	★★☆	Global	No	Frobenius norm	Transfer learning
45	MLR	2010	Yes	Full	Linear	★★☆	☆☆☆	Global	Yes	Nuclear norm	For ranking
47	LRML	2008	Yes	Semi	Linear	★☆☆	N/A	Global	No	Laplacian	—
47	M-DML	2009	No	Semi	Linear	★☆☆	N/A	Local	No	Laplacian	Auxiliary metrics
48	ARC-t	2011	No	Full	Linear	★☆☆	★☆☆	Global	No	Frobenius	Domain Adaptation
48	CDML	2011	No	Semi	N/A	N/A	N/A	N/A	N/A	N/A	Domain adaptation
49	DAML	2011	No	Semi	Nonlinear	N/A	N/A	Global	No	MMD	Domain adaptation
49	χ^2-LMNN	2012	No	Full	Nonlinear	★★☆	★★☆	Local	Yes	None	Histogram data
50	GML	2011	No	Weak	Linear	★☆☆	★★☆	Local	No	None	Histogram data
51	EMDL	2012	No	Weak	Linear	★☆☆	★★☆	Local	No	Frobenius norm	Histogram data

Table 10.2: Main features of metric learning methods for structured data.

Page	Name	Year	Source Code	Data Type	Method	Script	Optimum	Negative Pairs
53	R&Y	1998	Yes	String	Gen.+EM	All	Local	No
54	O&S	2006	Yes	String	Disc.+EM	All	Local	No
55	Saigo	2006	Yes	String	Gradient	All	Local	No
56	GESL	2011	Yes	String/tree/graph	Gradient	Levenshtein	Global	Yes
57	Bernard	2006	Yes	Tree	Both+EM	All	Local	No
57	Boyer	2007	Yes	Tree	Gen.+EM	All	Local	No
57	Dalvi	2009	No	Tree	Disc.+EM	All	Local	No
57	Emms	2012	No	Tree	Disc.+EM	Optimal	Local	No
57	N&B	2007	No	Graph	Gen.+EM	All	Local	No
58	MLTSA	2014	No	Time series	Gradient	N/A	Global	N/A

Leveraging the structure The simple example of metric learning designed specifically for histogram data [Kedem et al., 2012] has shown that taking the structure of the data into account when learning the metric can lead to significant improvements in performance. As data is becoming more and more structured, using this structure to guide the choice of metric is likely to receive increasing interest in the near future. An important case, is metric learning on graphs (e.g., social networks, knowledge bases) for link prediction, is a promising and under-explored topic. A preliminary work is due to Shaw et al. [2011]. Another example is to learn distances between matrices, which has applications for instance in computer vision (covariance region descriptors, medical imaging). In this case, one could consider a matrix as a large vector or compute distances based on distances between columns, but this ignores the 2D structure. A preliminary work on learning distances between PSD matrices is due to Vemulapalli and Jacobs [2015].

Adapting the metric to changing data An important issue is to develop methods robust to changes in the data. In this line of work, metric learning in the presence of noisy data as well as for transfer learning and domain adaptation has recently received some interest. However, these efforts are still insufficient for dealing with lifelong learning (or "learning to learn") applications, where the learner experiences concept drift and must detect and adapt the metric to changes in real time.

Unsupervised metric learning A natural question to ask is whether one can learn a metric in a purely unsupervised way. So far, this has only been done as a byproduct of dimensionality reduction algorithms. Other relevant criteria should be investigated, for instance learning a metric that is robust to noise or invariant to some transformations of interest, in the spirit of denoising autoencoders [Chen et al., 2012, Vincent et al., 2008]. Some results in this direction have been obtained for image transformations [Kumar et al., 2007]. A related problem is to characterize what it means for a metric to be good for clustering. There has been preliminary work on this question [Balcan et al., 2008b, Lajugie et al., 2014a], which probably deserves more attention.

Learning richer metrics The notion of similarity is often richer than what current metric learning algorithms can encode. For instance, there exist several ways in which two instances may be similar (perhaps based on different sets of features), and different degrees of similarity (versus the simple binary similar/dissimilar view). Being able to model these shades as well as to interpret what makes things similar would bring the learned metrics closer to the rich human notions of similarity.

More theoretical understanding Although several recent papers have looked at the generalization of metric learning, analyzing the link between the consistency of the learned metric and its performance in a given algorithm (classifier, clustering procedure, etc.) remains an important open problem. So far, only results for linear classification have been obtained [Bellet et al., 2012b, Guo and Ying, 2014], while learned metrics are also widely used in k-NN classification, clustering or information retrieval, for which no such theoretical result is known.

APPENDIX A

Proofs of Chapter 8

A.1 UNIFORM STABILITY

A.1.1 PROOF OF LEMMA 8.6

Proof. The first steps of this proof are similar to the proof of Lemma 20 in [Bousquet and Elisseeff, 2002] which we recall for the sake of completeness. Recall that any convex function g verifies

$$\forall x, y, \forall t \in [0,1], g(x + t(y-x)) - g(x) \leq t(g(x) - g(y)).$$

$\epsilon_{\mathcal{T}^{i,z}}(\cdot)$ is convex and thus for any $t \in [0,1]$,

$$\epsilon_{\mathcal{T}^{i,z}}(d_{\boldsymbol{M}_\mathcal{T} - t\Delta \boldsymbol{M}}) - \epsilon_{\mathcal{T}^{i,z}}(d_{\boldsymbol{M}_\mathcal{T}}) \leq t(\epsilon_{\mathcal{T}^{i,z}}(d_{\boldsymbol{M}_{\mathcal{T}^{i,z}}}) - \epsilon_{\mathcal{T}^{i,z}}(d_{\boldsymbol{M}_\mathcal{T}})). \tag{A.1}$$

Switching the role of $\boldsymbol{M}_\mathcal{T}$ and $\boldsymbol{M}_{\mathcal{T}^{i,z}}$, we get:

$$\epsilon_{\mathcal{T}^{i,z}}(d_{\boldsymbol{M}_{\mathcal{T}^{i,z}} + t\Delta \boldsymbol{M}}) - \epsilon_{\mathcal{T}^{i,z}}(d_{\boldsymbol{M}_{\mathcal{T}^{i,z}}}) \leq t(\epsilon_{\mathcal{T}^{i,z}}(d_{\boldsymbol{M}_\mathcal{T}}) - \epsilon_{\mathcal{T}^{i,z}}(d_{\boldsymbol{M}_{\mathcal{T}^{i,z}}})). \tag{A.2}$$

Summing up inequalities (A.1) and (A.2) yields

$$\epsilon_{\mathcal{T}^{i,z}}(d_{\boldsymbol{M}_\mathcal{T} - t\Delta \boldsymbol{M}}) - \epsilon_{\mathcal{T}^{i,z}}(\boldsymbol{M}_\mathcal{T}) + \epsilon_{\mathcal{T}^{i,z}}(d_{\boldsymbol{M}_{\mathcal{T}^{i,z}} + t\Delta \boldsymbol{M}}) - \epsilon_{\mathcal{T}^{i,z}}(d_{\boldsymbol{M}_{\mathcal{T}^{i,z}}}) \leq 0. \tag{A.3}$$

Now, since $\boldsymbol{M}_\mathcal{T}$ and $\boldsymbol{M}_{\mathcal{T}^{i,z}}$ are minimizers of $F_\mathcal{T}$ and $F_{\mathcal{T}^{i,z}}$ respectively, we have:

$$F_\mathcal{T}(d_{\boldsymbol{M}_\mathcal{T}}) - F_\mathcal{T}(d_{\boldsymbol{M}_\mathcal{T} - t\Delta \boldsymbol{M}}) \leq 0 \tag{A.4}$$
$$F_{\mathcal{T}^{i,z}}(d_{\boldsymbol{M}_{\mathcal{T}^{i,z}}}) - F_{\mathcal{T}^{i,z}}(d_{\boldsymbol{M}_{\mathcal{T}^{i,z}} + t\Delta \boldsymbol{M}}) \leq 0. \tag{A.5}$$

By summing up (A.4) and (A.5) we get:

$$\begin{aligned} &\epsilon_\mathcal{T}(d_{\boldsymbol{M}_\mathcal{T}}) + \lambda\|\boldsymbol{M}_\mathcal{T}\|_\mathcal{F} - (\epsilon_\mathcal{T}(d_{\boldsymbol{M}_\mathcal{T} - t\Delta \boldsymbol{M}}) + \lambda\|\boldsymbol{M}_\mathcal{T} - t\Delta \boldsymbol{M}\|_\mathcal{F}) + \\ &\epsilon_{\mathcal{T}^{i,z}}(d_{\boldsymbol{M}_{\mathcal{T}^{i,z}}}) + \lambda\|\boldsymbol{M}_{\mathcal{T}^{i,z}}\|_\mathcal{F} - (\epsilon_{\mathcal{T}^{i,z}}(d_{\boldsymbol{M}_{\mathcal{T}^{i,z}} + t\Delta \boldsymbol{M}}) + \lambda\|\boldsymbol{M}_{\mathcal{T}^{i,z}} + t\Delta \boldsymbol{M}\|_\mathcal{F}) \leq 0. \end{aligned}$$

By summing this last inequality with (A.3), we obtain

$$\begin{aligned} &\epsilon_\mathcal{T}(d_{\boldsymbol{M}_\mathcal{T}}) + \lambda\|\boldsymbol{M}_\mathcal{T}\|_\mathcal{F} - (\epsilon_\mathcal{T}(d_{\boldsymbol{M}_\mathcal{T} - t\Delta \boldsymbol{M}}) + \lambda\|\boldsymbol{M}_\mathcal{T} - t\Delta \boldsymbol{M}\|_\mathcal{F}) + \\ &\lambda\|\boldsymbol{M}_{\mathcal{T}^{i,z}}\|_\mathcal{F} - (\lambda\|\boldsymbol{M}_{\mathcal{T}^{i,z}} + t\Delta \boldsymbol{M}\|_\mathcal{F}) + \epsilon_{\mathcal{T}^{i,z}}(d_{\boldsymbol{M}_\mathcal{T} - t\Delta \boldsymbol{M}}) - \epsilon_{\mathcal{T}^{i,z}}(d_{\boldsymbol{M}_\mathcal{T}}) \leq 0. \end{aligned}$$

Let $B = \epsilon_\mathcal{T}(d_{\boldsymbol{M}_\mathcal{T} - t\Delta \boldsymbol{M}}) - \epsilon_{\mathcal{T}^{i,z}}(d_{\boldsymbol{M}_\mathcal{T} - t\Delta \boldsymbol{M}}) - (\epsilon_\mathcal{T}(d_{\boldsymbol{M}_\mathcal{T}}) - \epsilon_{\mathcal{T}^{i,z}}(d_{\boldsymbol{M}_\mathcal{T}}))$, we have then

$$\lambda(\|\boldsymbol{M}_\mathcal{T}\|_\mathcal{F} - \|\boldsymbol{M}_\mathcal{T} - t(\Delta \boldsymbol{M})\|_\mathcal{F} + \|\boldsymbol{M}_{\mathcal{T}^{i,z}}\|_\mathcal{F} - \|\boldsymbol{M}_{\mathcal{T}^{i,z}} + t(\Delta \boldsymbol{M})\|_\mathcal{F}) \leq B. \tag{A.6}$$

We now derive a bound for B. z^i_j denotes the jth example of sample $\mathcal{T}^{i,z}$.

$$
\begin{aligned}
B &\leq |\epsilon_{\mathcal{T}}(d_{M_{\mathcal{T}}-t\Delta M}) - \epsilon_{\mathcal{T}^{i,z}}(d_{M_{\mathcal{T}}-t\Delta M}) - (\epsilon_{\mathcal{T}}(d_{M_{\mathcal{T}}}) - \epsilon_{\mathcal{T}^{i,z}}(d_{M_{\mathcal{T}}}))| \\
&\leq \frac{1}{n^2}\left|\sum_{k=1}^{n}\sum_{j=1}^{n}\ell(d_{M_{\mathcal{T}}-t\Delta M}, z_k, z_j) - \sum_{k=1}^{n}\sum_{j=1}^{n}\ell(d_{M_{\mathcal{T}}-t\Delta M}, z^i_k, z^i_j)\right. \\
&\qquad \left. - \left(\sum_{k=1}^{n}\sum_{j=1}^{n}\ell(d_{M_{\mathcal{T}}}, z_k, z_j) - \sum_{k=1}^{n}\sum_{j=1}^{n}\ell(d_{M_{\mathcal{T}}}, z^i_k, z^i_j)\right)\right| \\
&\leq \frac{1}{n^2}\left|\sum_{j=1}^{n}\left(\ell(d_{M_{\mathcal{T}}-t\Delta M}, z_i, z_j) - \ell(d_{M_{\mathcal{T}}-t\Delta M}, z, z^i_j)\right) + \right. \\
&\qquad \sum_{\substack{k=1\\k\neq i}}^{n}\sum_{j=1}^{n}\left(\ell(d_{M_{\mathcal{T}}-t\Delta M}, z_k, z_j) - \ell(d_{M_{\mathcal{T}}-t\Delta M}, z^i_k, z^i_j)\right) \\
&\qquad \left. - \left(\sum_{k=1}^{n}\sum_{j=1}^{n}\ell(d_{M_{\mathcal{T}}}, z_k, z_j) - \sum_{k=1}^{n}\sum_{j=1}^{n}\ell(d_{M_{\mathcal{T}}}, z^i_k, z^i_j)\right)\right|
\end{aligned}
$$

This inequality is obtained by developing the sum of the first two terms of the second line. Note that the subsamples of $n-1$ elements $\mathcal{T}\backslash\{z_i\}$ and $\mathcal{T}^{i,z}\backslash\{z\}$ are the same and thus $z_k = z^i_k$ when $k \neq i$. Therefore, some terms cancel out and we have:

$$
\begin{aligned}
B \leq \frac{1}{n^2}&\left|\sum_{j=1}^{n}\left(\ell(d_{M_{\mathcal{T}}-t\Delta M}, z_i, z_j) - \ell(d_{M_{\mathcal{T}}-t\Delta M}, z, z^i_j)\right) + \sum_{\substack{k=1\\k\neq i}}^{n}\left(\ell(d_{M_{\mathcal{T}}-t\Delta M}, z_k, z_i)\right.\right. \\
&\left.\left. -\ell(d_{M_{\mathcal{T}}-t\Delta M}, z_k, z)\right) - \left(\sum_{k=1}^{n}\sum_{j=1}^{n}\ell(d_{M_{\mathcal{T}}}, z_k, z_j) - \sum_{k=1}^{n}\sum_{j=1}^{n}\ell(d_{M_{\mathcal{T}}}, z_k, z^i_j)\right)\right|.
\end{aligned}
$$

The first two sums in the absolute value can be bounded by:

$$
(2n-1) \sup_{\substack{z_1,z_2\in\mathcal{T}\\z_3,z_4\in\mathcal{T}^{i,z}}} |\ell(d_{M_{\mathcal{T}}-t\Delta M}, z_1, z_2) - \ell(d_{M_{\mathcal{T}}-t\Delta M}, z_3, z_4)|.
$$

The same analysis can be done for the part in parentheses of the last line of the absolute value and we can take the pair of examples in $\mathcal{T}$ and in $\mathcal{T}^{i,z}$ maximizing the whole absolute value to obtain the next inequality:

$$B \leq \frac{2n-1}{n^2} \sup_{\substack{z_1,z_2 \in \mathcal{T} \\ z_3,z_4 \in \mathcal{T}^{i,z}}} |\ell(d_{M_\mathcal{T}-t\Delta M}, z_1, z_2) - \ell(d_{M_\mathcal{T}-t\Delta M}, z_3, z_4) - \left(\ell(d_{M_\mathcal{T}}, z_1, z_2) - \ell(d_{M_\mathcal{T}}, z_3, z_4)\right)\Big| .$$

We continue by applying a reordering of the terms and the triangular inequality to get the next result:

$$B \leq \frac{2n-1}{n^2}\left(\sup_{z_1,z_2 \in \mathcal{T}} |\ell(d_{M_\mathcal{T}-t\Delta M}, z_1, z_2) - \ell(d_{M_\mathcal{T}}, z_1, z_2)| + \sup_{z_3,z_4 \in \mathcal{T}^{i,z}} |\ell(d_{M_\mathcal{T}-t\Delta M}, z_3, z_4) - \ell(d_{M_\mathcal{T}}, z_3, z_4)| \right).$$

We then use twice the k-lipschitz property of ℓ which leads to:

$$\begin{aligned} B &\leq \frac{(2n-1)}{n^2} 2k\| - t\Delta M\|_\mathcal{F} \\ &\leq \frac{(2n)}{n^2} 2kt\|\Delta M\|_\mathcal{F} = \frac{4tk}{n}\|\Delta M\|_\mathcal{F}. \end{aligned}$$

Then, by applying this bound on B from inequality (A.6), we get the lemma. □

A.1.2 PROOF OF THEOREM 8.7

Proof. Using $t = 1/2$ on the left-hand side of Lemma 8.6, we get

$$\|M_T\|_\mathcal{F}^2 - \|M_T - \frac{1}{2}\Delta M\|_\mathcal{F}^2 + \|M_{\mathcal{T}^{i,z}}\|_\mathcal{F}^2 - \|M_{\mathcal{T}^{i,z}} + \frac{1}{2}\Delta M\|_\mathcal{F}^2 = \frac{1}{2}\|\Delta M\|_\mathcal{F}^2.$$

Then, applying Lemma 8.6, we get

$$\frac{1}{2}\|\Delta M\|_\mathcal{F}^2 \leq \frac{2k}{\lambda n}\|\Delta M\|_\mathcal{F} \Rightarrow \|\Delta M\|_\mathcal{F} \leq \frac{4k}{\lambda n}.$$

Now, from the k-lipschitz property of ℓ, we have for any z, z'

$$|\ell(d^2_{M_\mathcal{T}}, z, z') - \ell(d^2_{M_{\mathcal{T}\rangle,\ddagger}}, z, z')| \leq k\|\Delta M\|_\mathcal{F} \leq \frac{4k^2}{\lambda n}.$$

□

A.1.3 PROOF OF LEMMA 8.9

Proof. First recall that for any $\mathcal{T}, z, z'$, by hypothesis of uniform stability we have:

$$\ell(d^2_{M_\mathcal{T}}, z, z') - \ell(d^2_{M_{\mathcal{T}^{k,z}}}, z, z')| \leq \sup_{z_1,z_2} |\ell(d^2_{M_\mathcal{T}}, z_1, z_2) - \ell(d^2_{M_{\mathcal{T}^{k,z}}}, z_1, z_2)| \leq \frac{\kappa}{n}.$$

Now, we can derive a bound for $\mathbf{E}_\mathcal{T}[D_\mathcal{T}]$.

$$\begin{aligned}
\mathbf{E}_\mathcal{T}[D_\mathcal{T}] &\leq \mathbf{E}_\mathcal{T}[\mathbf{E}_{z,z'}[\ell(d^2_{M_\mathcal{T}}, z, z')] - L_\mathcal{T}(M_\mathcal{T})] \\
&\leq \mathbf{E}_{\mathcal{T},z,z'}[|\ell(d^2_{M_\mathcal{T}}, z, z') - \frac{1}{n}\sum_{k=1}^{n}\frac{1}{n}\sum_{j=1}^{n}\ell(d^2_{M_\mathcal{T}}, z_k, z'_j)|] \\
&\leq \mathbf{E}_{\mathcal{T},z,z'}[|\frac{1}{n}\sum_{k=1}^{n}\frac{1}{n}\sum_{j=1}^{n}(\ell(d^2_{M_\mathcal{T}}, z, z') - \ell(d^2_{M_{\mathcal{T}^{k,z}}}, z_k, z'_j) + \\
&\qquad\qquad \ell(d^2_{M_{\mathcal{T}^{k,z}}}, z_k, z'_j) - \ell(d^2_{M_\mathcal{T}}, z_k, z'_j))|] \\
&\leq \mathbf{E}_{\mathcal{T},z,z'}[|\frac{1}{n}\sum_{k=1}^{n}\frac{1}{n}\sum_{j=1}^{n}(\ell(d^2_{M_\mathcal{T}}, z, z') - \ell(d^2_{M_{\mathcal{T}^{k,z}}}, z_k, z'_{k_j}))|] + \\
&\quad \frac{1}{n}\sum_{k=1}^{n}\frac{1}{n}\sum_{j=1}^{n}\mathbf{E}_{\mathcal{T},z,z'}[|\ell(d^2_{M_{\mathcal{T}^{k,z}}}, z_k, z'_j) - \ell(d^2_{M_\mathcal{T}}, z_k, z'_j)|] \\
&\leq \mathbf{E}_{\mathcal{T},z,z'}[|\frac{1}{n}\sum_{k=1}^{n}\frac{1}{n}\sum_{j=1}^{n}(\ell(d^2_{M_\mathcal{T}}, z, z') - \ell(d^2_{M_{\mathcal{T}^{k,z}}}, z_k, z'_j))|] + \frac{\kappa}{n}.
\end{aligned}$$

The last inequality is obtained by applying the hypothesis of uniform stability to the second part of the sum. Now, since $\mathcal{T}$, z and z' are i.i.d. from distribution μ, we do not change the expected value by replacing one point with another and thus:

$$\mathbf{E}_{\mathcal{T},z,z'}[|\ell(d^2_{M_\mathcal{T}}, z, z') - \ell(d^2_{M_\mathcal{T}}, z_k, z')|] = \mathbf{E}_{\mathcal{T},z,z'}[|\ell(d^2_{M^{z,k}_\mathcal{T}}, z_k, z') - \ell(d^2_{M_\mathcal{T}}, z_k, z')|].$$

Then we get the result by applying this trick twice on the first element of the sum:

$$\begin{aligned}
\mathbf{E}_\mathcal{T}[D_\mathcal{T}] &\leq \mathbf{E}_{\mathcal{T},z,z'}[|\frac{1}{n}\sum_{k=1}^{n}\frac{1}{n}\sum_{j=1}^{n}(\ell(d^2_{M_{\mathcal{T}^{k,z}}}, z_k, z') - \ell(d^2_{M_\mathcal{T}}, z_k, z'_j))|] + \frac{\kappa}{n} \\
&\leq \mathbf{E}_{\mathcal{T},z,z'}[|\frac{1}{n}\sum_{k=1}^{n}\frac{1}{n}\sum_{j=1}^{n}(\ell(d^2_{M_{\{\mathcal{T}^{k,z}\}^{j,z'}}}, z_k, z'_j) - \ell(d^2_{M_{\mathcal{T}^{k,z}}}, z_k, z'_j))|] + \frac{\kappa}{n} \\
&\leq \frac{\kappa}{n} + \frac{\kappa}{n}.
\end{aligned}$$

□

A.1.4 PROOF OF LEMMA 8.10

Proof. First, we derive a bound on $|D_{\mathcal{T}} - D_{\mathcal{T}^{i,z}}|$.

$$
\begin{aligned}
&|D_{\mathcal{T}} - D_{\mathcal{T}^{i,z}}| \\
&= |\epsilon(d^2_{M_\mathcal{T}}) - \epsilon_{\mathcal{T}}(d^2_{M_\mathcal{T}}) - (\epsilon(d^2_{M_{\mathcal{T}^{i,z}}}) - \epsilon_{\mathcal{T}^{i,z}}(d^2_{M_{\mathcal{T}^{i,z}}}))| \\
&= |\epsilon(d^2_{M_\mathcal{T}}) - \epsilon_{\mathcal{T}}(d^2_{M_\mathcal{T}}) - \epsilon(d^2_{M_{\mathcal{T}^{i,z}}}) + \epsilon_{\mathcal{T}^{i,z}}(d^2_{M_{\mathcal{T}^{i,z}}}) + \epsilon_{\mathcal{T}}(d^2_{M_{\mathcal{T}^{i,z}}}) - \epsilon_{\mathcal{T}}(d^2_{M_{\mathcal{T}^{i,z}}})| \\
&= |\epsilon(d^2_{M_\mathcal{T}}) - \epsilon(d^2_{M_{\mathcal{T}^{i,z}}}) + \epsilon_{\mathcal{T}}(d^2_{M_{\mathcal{T}^{i,z}}}) - \epsilon_{\mathcal{T}}(d^2_{M_\mathcal{T}}) + \epsilon_{\mathcal{T}^{i,z}}(d^2_{M_{\mathcal{T}^{i,z}}}) - \epsilon_{\mathcal{T}}(d^2_{M_{\mathcal{T}^{i,z}}})| \\
&\leq |\epsilon(d^2_{M_\mathcal{T}}) - \epsilon(d^2_{M_{\mathcal{T}^{i,z}}})| + |\epsilon_{\mathcal{T}}(d^2_{M_{\mathcal{T}^{i,z}}}) - \epsilon_{\mathcal{T}}(d^2_{M_\mathcal{T}})| + |\epsilon_{\mathcal{T}^{i,z}}(d^2_{M_{\mathcal{T}^{i,z}}}) - \epsilon_{\mathcal{T}}(d^2_{M_{\mathcal{T}^{i,z}}})| \\
&\leq \mathbf{E}_{z_1,z_2}[|\ell(d^2_{M_\mathcal{T}}, z_1, z_2) - \ell(d^2_{M_{\mathcal{T}^{i,z}}}, z_1, z_2)|] + \\
&\quad \frac{1}{n}\sum_{k=1}^{n}\frac{1}{n}\sum_{j=1}^{n}|\ell(d^2_{M_{\mathcal{T}^{i,z}}}, z_k, z'_j) - \ell(d^2_{M_\mathcal{T}}, z_k, z'_j)| + |\epsilon_{\mathcal{T}^{i,z}}(d^2_{M_{\mathcal{T}^{i,z}}}) - \epsilon_{\mathcal{T}}(d^2_{M_{\mathcal{T}^{i,z}}})| \\
&\leq 2\frac{\kappa}{n} + |\epsilon_{\mathcal{T}^{i,z}}(d^2_{M_{\mathcal{T}^{i,z}}}) - \epsilon_{\mathcal{T}}(d^2_{M_{\mathcal{T}^{i,z}}})| \text{ by using the hypothesis of stability twice.}
\end{aligned}
$$

Now, proving Lemma 8.10 boils down to bounding the last term above. Using arguments similar to those used in the second part of the proof of Lemma 8.6, we get

$$
|\epsilon_{\mathcal{T}^{i,z}}(d^2_{M_{\mathcal{T}^{i,z}}}) - \epsilon_{\mathcal{T}}(d^2_{M_{\mathcal{T}^{i,z}}})| \leq \frac{(2)}{n} \sup_{\substack{z_1,z_2\in\mathcal{T} \\ z_3,z_4\in\mathcal{T}^{i,z}}} |\ell(d^2_{M_{\mathcal{T}^{i,z}}}, z_1, z_2) - \ell(d^2_{M_{\mathcal{T}^{i,z}}}, z_3, z_4)|.
$$

Now by the (σ, m)-admissibility of ℓ, we have that:

$$
|\ell(d^2_{M_{\mathcal{T}^{i,z}}}, z_1, z_2) - \ell(d^2_{M_{\mathcal{T}^{i,z}}}, z_3, z_4)| \leq \sigma(|y_{12} - y_{34}| + m \leq 2\sigma + m,
$$

since whatever the labels, $|y_{12} - y_{34}| \leq 2$. This leads us to the desired result. □

A.1.5 PROOF OF LEMMA 8.12

Proof. If $M_\mathcal{T}$ is an optimal solution, then the value reached by the objective function is lower than the one obtained with the null matrix $\mathbf{0}$:

$$
\frac{1}{n^2}\sum_{z_i,z_j\in\mathcal{T}} \ell(d^2_M, z_i, z_j) + \lambda\|M_\mathcal{T}\|^2_\mathcal{F} \leq \frac{1}{n^2}\sum_{z_i,z_j\in\mathcal{T}} \ell(d^2_\mathbf{0}, z_i, z_j) + \lambda\|\mathbf{0}_\mathcal{T}\|^2_\mathcal{F} \leq c.
$$

For the last inequality, note that regardless of the possible labels of z_i and z_j, $\ell(d^2_\mathbf{0}, z_i, z_j)$ is bounded either by c or 0.
Since $\frac{1}{n^2}\sum_{z_i,z_j\in\mathcal{T}} \ell(d^2_M, z_i, z_j)$, we get $\lambda\|M_\mathcal{T}\|^2_\mathcal{F} \leq c$. □

A.1.6 PROOF OF LEMMA 8.13

Proof. We need to bound $|\ell(d^2_M, z_i, z_j) - \ell(d^2_{\mathbf{M}'}, z_i, z_j)|$.

$$\begin{aligned}
|\ell(d^2_M, z_i, z_j) - \ell(d^2_{\mathbf{M}'}, z_i, z_j)| &\le |y_{ij}[c - d^2_M(\boldsymbol{x}_i, \boldsymbol{x}_j)] - y_{ij}[c - d^2_{M'}(\boldsymbol{x}_i, \boldsymbol{x}_j)]| \\
&\le |d^2_{M'}(\boldsymbol{x}_i, \boldsymbol{x}_j) - d^2_M(\boldsymbol{x}_i, \boldsymbol{x}_j)| = |d^2_M(\boldsymbol{x}_i, \boldsymbol{x}_j) - d^2_{M'}(\boldsymbol{x}_i, \boldsymbol{x}_j)| \\
&\le |(\boldsymbol{x}_i - \boldsymbol{x}_j)^T(\boldsymbol{M} - \boldsymbol{M}')(\boldsymbol{x}_i - \boldsymbol{x}_j)| \\
&\le \|(\boldsymbol{x}_i - \boldsymbol{x}_j)\| \|(\boldsymbol{M} - \boldsymbol{M}')(\boldsymbol{x}_i - \boldsymbol{x}_j)\| \\
&\le \|(\boldsymbol{x}_i - \boldsymbol{x}_j)\| \|(\boldsymbol{M} - \boldsymbol{M}')\| \|(\boldsymbol{x}_i - \boldsymbol{x}_j)\| \\
&\le 4B^2 \|(\boldsymbol{M} - \boldsymbol{M}')\|.
\end{aligned}$$

The first inequality is obtained by the 1-lipschitz property of the hinge loss:

$$|[X]_+ - [Y]_+| \le |X - Y|.$$

The last inequality is obtained by the fact that each instance $\boldsymbol{x}$ is B-bounded: $\|\boldsymbol{x}\| \le B$. □

A.1.7 PROOF OF LEMMA 8.14

Proof. First, since the hinge loss is 1-lipschitz we get:

$$\begin{aligned}
|\ell(d^2_M, z_1, z_2) - \ell(d^2_M, z_3, z_4)| &\le |y_{12}[c - d^2_M(\boldsymbol{x}_1, \boldsymbol{x}_2)] - y_{34}[c - d^2_M(\boldsymbol{x}_3, \boldsymbol{x}_4)]| \\
&\le |c(y_{12} - y_{34}) + |y_{34} d^2_M(\boldsymbol{x}_3, \boldsymbol{x}_4) - y_{12} d^2_M(\boldsymbol{x}_1, \boldsymbol{x}_2)| \\
&\le c|y_{12} - y_{34}| + 2 \times 4B^2 \frac{c}{\lambda}.
\end{aligned}$$

For the last inequality, whatever the values of y_{12} and y_{34}, the right-hand side term of the previous inequality is bounded by $|d^2_M(\boldsymbol{x}_3, \boldsymbol{x}_4)| + |d^2_M(\boldsymbol{x}_1, \boldsymbol{x}_2)|$. Now considering that for any $\boldsymbol{x}_i$ and $\boldsymbol{x}_j$, $\|\boldsymbol{x}_i - \boldsymbol{x}_j\| \le 2B$ and by applying the Cauchy-Schwarz inequality, classical result on norms and Lemma 8.12, we can get that
$|d^2_M(\boldsymbol{x}_3, \boldsymbol{x}_4)| + |d^2_M(\boldsymbol{x}_1, \boldsymbol{x}_2)| \le 2 \times 4B^2 \frac{c}{\lambda}$.
Taking $\sigma = c$ and $m = 8B^2 \frac{c}{\lambda}$ proves the lemma. □

A.2 ALGORITHMIC ROBUSTNESS

A.2.1 PRELIMINARY RESULT

We need the following concentration inequality that will help us to derive the bound.

Proposition A.1 van der Vaart and Wellner [2000] *Let $(|N_1|, \ldots, |N_K|)$ an i.i.d. multinomial random variable with parameters n and $(\mu(C_1), \ldots, \mu(C_K))$. By the Bretagnolle-Huber-Carol inequality we have:*

$$Pr\left\{\sum_{i=1}^{K} \left|\frac{|N_i|}{n} - \mu(C_i)\right| \ge \lambda\right\} \le 2^K \exp\left(\frac{-n\lambda^2}{2}\right).$$

Hence, with probability at least $1-\delta$,

$$\sum_{i=1}^{K}\left|\frac{N_i}{n}-\mu(C_i)\right| \leq \sqrt{\frac{2K\ln 2+2\ln(1/\delta)}{n}}. \tag{A.7}$$

A.2.2 PROOF OF THEOREM 8.17

Proof. Let N_i be the set of index of points of $\mathcal{T}$ that fall into the C_i. $(|N_1|,\dots,|N_K|)$ is an i.i.d. random variable with parameters n and $(\mu(C_1),\dots,\mu(C_K))$. We have:

$$\begin{aligned}
&|\epsilon(d^2_{M_\mathcal{T}})-\epsilon_\mathcal{T}(d^2_{M_\mathcal{T}})|\\
=\;& \left|\sum_{i,j=1}^{K}\mathbb{E}_{z_1',z_2'\sim\mu}\left(\ell(d^2_{M_\mathcal{T}},z_1',z_2')|z_1'\in C_i,z_2'\in C_j\right)\mu(C_i)\mu(C_j)-\frac{1}{n^2}\sum_{i,j=1}^{n}\ell(d^2_{M_\mathcal{T}},z_i,z_j)\right|\\
\stackrel{(a)}{\leq}\;& \left|\sum_{i,j=1}^{K}\mathbb{E}_{z_1',z_2'\sim\mu}\left(\ell(d^2_{M_\mathcal{T}},z_1',z_2')|z_1'\in C_i,z_2'\in C_j\right)\mu(C_i)\mu(C_j)-\right.\\
&\qquad\left.\sum_{i,j=1}^{K}\mathbb{E}_{z_1',z_2'\sim\mu}\left(\ell(d^2_{M_\mathcal{T}},z_1',z_2')|z_1'\in C_i,z_2'\in C_j\right)\mu(C_i)\frac{|N_j|}{n}\right|+\\
&\left|\sum_{i,j=1}^{K}\mathbb{E}_{z_1',z_2'\sim\mu}\left(\ell(d^2_{M_\mathcal{T}},z_1',z_2')|z_1'\in C_i,z_2'\in C_j\right)\mu(C_i)\frac{|N_j|}{n}-\frac{1}{n^2}\sum_{i,j=1}^{n}l(d^2_{M_\mathcal{T}},z_i,z_j)\right|\\
\stackrel{(b)}{\leq}\;& \left|\sum_{i,j=1}^{K}\mathbb{E}_{z_1',z_2'\sim\mu}\left(\ell(d^2_{M_\mathcal{T}},z_1',z_2')|z_1'\in C_i,z_2'\in C_j\right)\mu(C_i)(\mu(C_j)-\frac{|N_j|}{n})\right|+\\
&\left|\sum_{i,j=1}^{K}\mathbb{E}_{z_1',z_2'\sim\mu}\left(\ell(d^2_{M_\mathcal{T}},z_1',z_2')|z_1'\in C_i,z_2'\in C_j\right)\mu(C_i)\frac{|N_j|}{n}-\right.\\
&\qquad\left.\sum_{i,j=1}^{K}\mathbb{E}_{z_1',z_2'\sim\mu}\left(\ell(d^2_{M_\mathcal{T}}),z_1',z_2')|z_1'\in C_i,z_2'\in C_j\right)\frac{|N_i||N_j|}{n}\right|+\\
&\left|\sum_{i,j=1}^{K}\mathbb{E}_{z_1',z_2'\sim\mu}\left(\ell(d^2_{M_\mathcal{T}}),z_1',z_2')|z_1'\in C_i,z_2'\in C_j\right)\frac{|N_i||N_j|}{n}-\frac{1}{n^2}\sum_{i,j=1}^{n}\ell(d^2_{M_\mathcal{T}}),z_i,z_j)\right|\\
\stackrel{(c)}{\leq}\;& B\left(\left|\sum_{j=1}^{K}\mu(C_j)-\frac{|N_j|}{n}\right|+\left|\sum_{i=1}^{K}\mu(C_i)-\frac{|N_i|}{n}\right|\right)+\\
&\left|\frac{1}{n^2}\sum_{i,j=1}^{K}\sum_{z_o\in N_i}\sum_{z_l\in N_j}\max_{z\in C_i}\max_{z'\in C_j}|\ell(d^2_{M_\mathcal{T}},z,z')-\ell(d^2_{M_\mathcal{T}},z_o,z_l)|\right|
\end{aligned}$$

$$\overset{(d)}{\leq} \omega(\mathcal{P}(\mathcal{T})) + 2B \sum_{i=1}^{K} \left| \frac{|N_i|}{n} - \mu(C_i) \right| \overset{(e)}{\leq} \omega(\mathcal{P}(\mathcal{T})) + 2B \sqrt{\frac{2K \ln 2 + 2 \ln 1/\delta}{n}}.$$

Inequalities (a) and (b) are due to the triangle inequality, (c) uses the fact that ℓ is bounded by B, that $\sum_{i=1}^{K} \mu(C_i) = 1$ by definition of a multinomial random variable and that $\sum_{j=1}^{K} \frac{|N_j|}{n} = 1$ by definition of the N_j. Lastly, (d) is due to the hypothesis of robustness (Equation 8.3) and (e) to the application of Proposition A.1. □

A.2.3 PROOF OF THEOREM 8.20

Proof. By definition of covering number, we can partition $\mathcal{X}$ in $\mathcal{N}(\gamma/2, \mathcal{X}, \rho)$ subsets such that each subset has a diameter less or equal to γ. Furthermore, since Y is a finite set, we can partition $\mathcal{Z}$ into $|Y|\mathcal{N}(\gamma/2, \mathcal{X}, \rho)$ subsets $\{C_i\}$ such that $z_1, z_1' \in C_i \Rightarrow \rho(z_1, z_1') \leq \gamma$ and $y_1 = y_1'$. Therefore, $\forall z_1, z_2, z_1', z_2' : z_1, z_2 \in \mathcal{T}, \rho(z_1, z_1') \leq \gamma, \rho(z_2, z_2') \leq \gamma$,

$$|\ell(d_M^2, z_1, z_2) - \ell(d_M^2, z_1', z_2')| \leq \omega(\mathcal{P}(\mathcal{T})),$$

this implies:
$z_1, z_2 \in \mathbf{s}, z_1, z_1' \in C_i, z_2, z_2' \in C_j \Rightarrow |\ell(d_M^2, z_1, z_2) - \ell(d_M^2, z_1', z_2')| \leq \omega(\mathcal{P}(\mathcal{T}))$,
which establishes the theorem. □

A.2.4 PROOF OF EXAMPLE 8.22

Proof. Let $\mathbf{M}^*$ be the solution given training sample $\mathcal{T}$. Due to the optimality of $\mathbf{M}^*$, we have $\|\mathbf{M}^*\|_1 \leq g_0/\lambda$. We can partition $\mathcal{Z}$ as $|Y|\mathcal{N}(\gamma/2, \mathcal{X}, \|\cdot\|_1)$ sets, such that if z and z' belong to the same set, then $y = y'$ and $\|\boldsymbol{x} - \boldsymbol{x}'\|_1 \leq \gamma$. Now, for $z_1, z_2, z_1', z_2' \in \mathcal{Z}$, if $y_1 = y_1'$, $\|\boldsymbol{x}_1 - \boldsymbol{x}_1'\|_1 \leq \gamma$, $y_2 = y_2'$ and $\|\boldsymbol{x}_2 - \boldsymbol{x}_2'\|_1 \leq \gamma$, then:

$$\begin{aligned}
&|g(y_{12}[1 - d_{\mathbf{M}^*}^2(\boldsymbol{x}_1, \boldsymbol{x}_2)]) - g(y_{12}'[1 - d_{\mathbf{M}^*}^2(\boldsymbol{x}_1', \boldsymbol{x}_2')])| \\
&\quad\leq U(|(\boldsymbol{x}_1 - \boldsymbol{x}_2)^T \mathbf{M}^* (\boldsymbol{x}_1 - \boldsymbol{x}_1')| + |(\boldsymbol{x}_1 - \boldsymbol{x}_2)^T \mathbf{M}^* (\boldsymbol{x}_2' - \boldsymbol{x}_2)| \\
&\qquad + |(\boldsymbol{x}_1 - \boldsymbol{x}_1')^T \mathbf{M}^* (\boldsymbol{x}_1' + \boldsymbol{x}_2')| + |(\boldsymbol{x}_2' - \boldsymbol{x}_2)^T \mathbf{M}^* (\boldsymbol{x}_1' + \boldsymbol{x}_2')|) \\
&\quad\leq U(\|\boldsymbol{x}_1 - \boldsymbol{x}_2\|_\infty \|\mathbf{M}^*\|_1 \|\boldsymbol{x}_1 - \boldsymbol{x}_1'\|_1 + \|\boldsymbol{x}_1 - \boldsymbol{x}_2\|_\infty \|\mathbf{M}^*\|_1 \|\boldsymbol{x}_2' - \boldsymbol{x}_2\|_1 \\
&\qquad + \|\boldsymbol{x}_1 - \boldsymbol{x}_1'\|_1 \|\mathbf{M}^*\|_1 \|\boldsymbol{x}_1' - \boldsymbol{x}_2'\|_\infty + \|\boldsymbol{x}_2' - \boldsymbol{x}_2\|_1 \|\mathbf{M}^*\|_1 \|\boldsymbol{x}_1' - \boldsymbol{x}_2'\|_\infty) \\
&\quad\leq \frac{8UR\gamma g_0}{\lambda}.
\end{aligned}$$

□

A.2.5 PROOF OF EXAMPLE 8.23

Proof. Let $\mathbf{M}^*$ be the solution given training sample $\mathcal{T}$. Due to optimality of $\mathbf{M}^*$, we have $\|\mathbf{M}^*\|_{2,1} \leq g_0/\lambda$. We can partition $\mathcal{Z}$ in the same way as in the proof of Example 8.21 and use the inequality $\|\mathbf{M}^*\|_{\mathcal{F}} \leq \|\mathbf{M}^*\|_{2,1}$ (from Theorem 3 of Feng [2003], Klaus and Li [1995]) to derive the same bound:

$$\begin{aligned}
&|g(y_{12}[1 - d^2_{\mathbf{M}^*}(\boldsymbol{x}_1, \boldsymbol{x}_2)]) - g(y'_{12}[1 - d^2_{\mathbf{M}^*}(\boldsymbol{x}'_1, \boldsymbol{x}'_2)])| \\
\leq\ & U(\|\boldsymbol{x}_1 - \boldsymbol{x}_2\|_2 \|\mathbf{M}^*\|_{\mathcal{F}} \|\boldsymbol{x}_1 - \boldsymbol{x}'_1\|_2 + \|\boldsymbol{x}_1 - \boldsymbol{x}_2\|_2 \|\mathbf{M}^*\|_{\mathcal{F}} \|\boldsymbol{x}'_2 - \boldsymbol{x}_2\|_2 \\
&+ \|\boldsymbol{x}_1 - \boldsymbol{x}'_1\|_2 \|\mathbf{M}^*\|_{\mathcal{F}} \|\boldsymbol{x}'_1 - \boldsymbol{x}'_2\|_2 + \|\boldsymbol{x}'_2 - \boldsymbol{x}_2\|_2 \|\mathbf{M}^*\|_{\mathcal{F}} \|\boldsymbol{x}'_1 - \boldsymbol{x}'_2\|_2) \\
\leq\ & U(\|\boldsymbol{x}_1 - \boldsymbol{x}_2\|_2 \|\mathbf{M}^*\|_{2,1} \|\boldsymbol{x}_1 - \boldsymbol{x}'_1\|_2 + \|\boldsymbol{x}_1 - \boldsymbol{x}_2\|_2 \|\mathbf{M}^*\|_{2,1} \|\boldsymbol{x}'_2 - \boldsymbol{x}_2\|_2 \\
&+ \|\boldsymbol{x}_1 - \boldsymbol{x}'_1\|_2 \|\mathbf{M}^*\|_{2,1} \|\boldsymbol{x}'_1 - \boldsymbol{x}'_2\|_2 + \|\boldsymbol{x}'_2 - \boldsymbol{x}_2\|_2 \|\mathbf{M}^*\|_{2,1} \|\boldsymbol{x}'_1 - \boldsymbol{x}'_2\|_2) \\
\leq\ & \frac{8UR\gamma g_0}{\lambda}.
\end{aligned}$$

For the trace norm, we use the classical result $\|\mathbf{M}^*\|_{\mathcal{F}} \leq \|M\|_*$, which allows us to prove the same result by replacing $\|\cdot\|_{2,1}$ by $\|\cdot\|_*$ in the proof above. □

A.2.6 PROOF OF EXAMPLE 8.24

Proof. We assume $\mathbb{H}$ to be a Hilbert space with an inner product operator $\langle \cdot, \cdot \rangle$. The mapping $\phi(\cdot)$ is continuous from $\mathcal{X}$ to $\mathbb{H}$. The norm $\|\cdot\|_{\mathbb{H}} : \mathbb{H} \to \mathbb{R}$ is defined as $\|w\|_{\mathbb{H}} = \sqrt{\langle w, w \rangle}$ for all $w \in \mathbb{H}$. For matrices $\|M\|_{\mathbb{H}}$ we take the entry wise norm by considering a matrix as a vector, corresponding to the Frobenius norm. The kernel function is defined as $k(\boldsymbol{x}_1, \boldsymbol{x}_2) = \langle \phi(\boldsymbol{x}_1), \phi(\boldsymbol{x}_2) \rangle$.

B_γ and $f_{\mathbb{H}}(\gamma)$ are finite by the compactness of $\mathcal{X}$ and continuity of $k(\cdot, \cdot)$. Let $\mathbf{M}^*$ be the solution given training sample $\mathcal{T}$, by the optimality of $\mathbf{M}^*$ and using the same trick as the other examples we have: $\|\mathbf{M}^*\|_{\mathbb{H}} \leq g_0/c$. Then, by considering a partition of $\mathcal{Z}$ into $|Y|\mathcal{N}(\gamma/2, \mathcal{X}, \|\cdot\|_2)$ disjoint subsets such that if $(\boldsymbol{x}_1, y_1)$ and $(\boldsymbol{x}_2, y_2)$ belong to the same set then $y_1 = y_2$ and $\|\boldsymbol{x}_1 - \boldsymbol{x}_2\|_2 \leq \gamma$. We have then,

$$\begin{aligned}
&|g(y_{ij}[1 - f(\mathbf{M}^*, \phi(\boldsymbol{x}_1), \phi(\boldsymbol{x}_2))]) - g(y_{ij}[1 - f(\mathbf{M}^*, \phi(\boldsymbol{x}'_1), \phi(\boldsymbol{x}'_2))])| \\
\leq\ & U(|(\phi(\boldsymbol{x}_1) - \phi(\boldsymbol{x}_2))^T \mathbf{M}^* (\phi(\boldsymbol{x}_1) - \phi(\boldsymbol{x}'_1))| + \\
& |(\phi(\boldsymbol{x}_1) - \phi(\boldsymbol{x}_2))^T \mathbf{M}^* (\phi(\boldsymbol{x}'_2) - \phi(\boldsymbol{x}_2))| + \\
& |(\phi(\boldsymbol{x}_1) - \phi(\boldsymbol{x}'_1))^T \mathbf{M}^* (\phi(\boldsymbol{x}'_1) + \phi(\boldsymbol{x}'_2))| + \\
& |(\phi(\boldsymbol{x}'_2) - \phi(\boldsymbol{x}_2))^T \mathbf{M}^* (\phi(\boldsymbol{x}'_1) + \phi(\boldsymbol{x}'_2))|) \\
\leq\ & U(|\phi(\boldsymbol{x}_1)^T \mathbf{M}^* (\phi(\boldsymbol{x}_1) - \phi(\boldsymbol{x}'_1))| + |\phi(\boldsymbol{x}_2)^T \mathbf{M}^* (\phi(\boldsymbol{x}_1) - \phi(\boldsymbol{x}'_1))| + \\
& |\phi(\boldsymbol{x}_1)^T \mathbf{M}^* (\phi(\boldsymbol{x}'_2) \phi(\boldsymbol{x}_2))| + |\phi(\boldsymbol{x}_2)^T \mathbf{M}^* (\phi(\boldsymbol{x}'_2) - \phi(\boldsymbol{x}_2))| + \\
& |(\phi(\boldsymbol{x}_1) - \phi(\boldsymbol{x}'_1))^T \mathbf{M}^* \phi(\boldsymbol{x}'_1)| + |(\phi(\boldsymbol{x}_1) - \phi(\boldsymbol{x}'_1))^T \mathbf{M}^* \phi(\boldsymbol{x}'_2)| + \\
& |(\phi(\boldsymbol{x}'_2) - \phi(\boldsymbol{x}_2))^T \mathbf{M}^* \phi(\boldsymbol{x}'_1)| + |(\phi(\boldsymbol{x}'_2) - \phi(\boldsymbol{x}_2))^T \mathbf{M}^* \phi(\boldsymbol{x}'_2)|). \qquad \text{(A.8)}
\end{aligned}$$

Then, note that

$$\begin{aligned}
|\phi(\boldsymbol{x}_1)^T \mathbf{M}^* (\phi(\boldsymbol{x}_1) - \phi(\boldsymbol{x}'_1))| &\leq \sqrt{\langle \phi(\boldsymbol{x}_1), \phi(\boldsymbol{x}_1) \rangle} \|M^*\|_{\mathbb{H}} \sqrt{\langle \phi(\boldsymbol{x}'_1) - \phi(\boldsymbol{x}'_2), \phi(\boldsymbol{x}'_1) - \phi(\boldsymbol{x}'_2) \rangle} \\
&\leq B_\gamma \frac{g_o}{\lambda} \sqrt{f_{\mathbb{H}}(\gamma)}.
\end{aligned}$$

Thus, by applying the same principle to all the terms in the right part of inequality (A.8), we obtain:

$$|g(y_{ij}[1 - d^2_{\mathbf{M}^*}(\phi(\boldsymbol{x}_1), \phi(\boldsymbol{x}_2))]) - g(y_{ij}[1 - d^2_{\mathbf{M}^*}(\phi(\boldsymbol{x}'_1), \phi(\boldsymbol{x}'_2))])| \leq \frac{8UB_\gamma \sqrt{f_{\mathbb{H}}(\gamma)} g_0}{\lambda}.$$

□

A.3 SIMILARITY-BASED LINEAR CLASSIFIERS

A.3.1 PROOF OF LEMMA 8.31

Proof. $P1$ comes from $|K_M(\mathbf{x},\mathbf{x}')| \leq \|\mathbf{x}\|_2\|M\|_{\mathcal{F}}\|\mathbf{x}'\|_2$, the normalization on examples ($\|\mathbf{x}\|_2 \leq 1$) and the requirement on matrices ($\|M\|_{\mathcal{F}} \leq 1$).

For $P2$, $|K_M(\mathbf{x},\mathbf{x}') - K_{M'}(\mathbf{x},\mathbf{x}')| = |K_{M-M'}(\mathbf{x},\mathbf{x}')|$, and we use the normalization $\|\mathbf{x}\|_2 \leq 1$.

$P3$ follows directly from $|y| = 1$ and the 1-lipschitz property of the hinge loss: $|[X]_+ - [Y]_+| \leq |X - Y|$. □

A.3.2 PROOF OF LEMMA 8.32

Proof. We follow closely the beginning of the proof of 8.6, omitting some details to lighten the proof. Let $\Delta M = \mathbf{A^i} - M$ and $0 \leq t \leq 1$, $M_1 = \|M\|^2_{\mathcal{F}} - \|M + t\Delta M\|^2_{\mathcal{F}} + \|M^i\|^2_{\mathcal{F}} - \|M^i - t\Delta M\|^2_{\mathcal{F}}$ and $M_2 = \frac{1}{\lambda N_T}(\epsilon_{\mathcal{T}}(M_R) - \epsilon_{\mathcal{T}}((M + t\Delta M)_R) + \epsilon_{\mathcal{T}^i}((M + t\Delta M)_R) - \epsilon_{\mathcal{T}^i}(M_R))$. Using the fact that $F_{\mathcal{T}}$ and $F_{\mathcal{T}^i}$ are convex functions, that M and M^i are their respective minimizers and property P3, we have: $M_1 \leq M_2$. Fixing $t = 1/2$, we obtain $M_1 = \|M - M^i\|^2_{\mathcal{F}}$, and using property $P3$ and the normalization $\|\mathbf{x}\|_2 \leq 1$, we get:

$$M_2 \leq \frac{1}{\lambda N_T \gamma}(\|\frac{1}{2}\Delta M\|_{\mathcal{F}} + \| - \frac{1}{2}\Delta M\|_{\mathcal{F}}) = \frac{\|M - M^i\|_{\mathcal{F}}}{\lambda N_T \gamma}.$$

This leads to the inequality $\|M - M^i\|^2_{\mathcal{F}} \leq \frac{\|M-M^i\|_{\mathcal{F}}}{\lambda N_T \gamma}$ from which Lemma 8.32 is directly derived. □

We now have all the material needed to prove the stability property of our algorithm.

Lemma A.2 *Let n and n_R be the number of training examples and reasonable points respectively, $n_R = \hat{\tau} n_T$ with $\hat{\tau} \in]0, 1]$. SLLC has a uniform stability in $\frac{\kappa}{n}$ with $\kappa = \frac{1}{\gamma}(\frac{1}{\lambda\gamma} + \frac{2}{\hat{\tau}}) = \frac{\hat{\tau}+2\lambda\gamma}{\hat{\tau}\lambda\gamma^2}$, where λ is the regularization parameter and γ the margin.*

Proof. For any sample $\mathcal{T}$ of size n, any $1 \leq i \leq n$, any labeled examples $\mathbf{z} = (\mathbf{x}, y)$ and $\mathbf{z}'_i = (\mathbf{x}_i, y'_i) \sim \mu$:

$$
\begin{aligned}
&|\ell(\boldsymbol{M}, \mathbf{z}, R) - \ell(\boldsymbol{M}^i, \mathbf{z}, R^i)| \\
&\leq \left| \frac{1}{\gamma n_R} \sum_{k=1}^{n_R} y_k K_{\boldsymbol{M}}(\mathbf{x}, \mathbf{x}_k) - \frac{1}{\gamma N_{R^i}} \sum_{k=1}^{N_{R^i}} y_k K_{\boldsymbol{M}^i}(\mathbf{x}, \mathbf{x}_k) \right| \\
&= \left| \frac{1}{\gamma n_R} \left(\left(\sum_{k=1,k\neq i}^{n_R} (y_k K_{\boldsymbol{M}}(\mathbf{x}, \mathbf{x}_k) - K_{\boldsymbol{M}^i}(\mathbf{x}, \mathbf{x}_k)) \right) + y_i K_{\boldsymbol{M}}(\mathbf{x}, \mathbf{x}_i) - y'_i K_{\boldsymbol{M}^i}(\mathbf{x}, \mathbf{x}'_i) \right) \right| \\
&\leq \frac{1}{\gamma n_R} \left(\left(\sum_{k=1,k\neq i}^{n_R} (|y_k| \|\boldsymbol{M} - \boldsymbol{M}^i\|_{\mathcal{F}}) \right) + |y_i K_{\boldsymbol{M}^i}(\mathbf{x}, \mathbf{x}_i)| + |y'_i K_{\boldsymbol{M}}(\mathbf{x}, \mathbf{x'}_i)| \right) \\
&\leq \frac{1}{\gamma n_R} \left(\frac{n_R - 1}{\lambda n \gamma} + 2 \right) \\
&\leq \frac{1}{\gamma n_R} \left(\frac{n_R}{\lambda n \gamma} + 2 \right).
\end{aligned}
$$

The first inequality follows from $P3$. The second comes from the fact that R and R^i differ in at most one element, corresponding to the example $\mathbf{z}_i$ in R and the example $\mathbf{z}'_i$ replacing $\mathbf{z}_i$ in R^i. The last inequalities are obtained by the use of the triangle inequality, $P1$, $P2$, Lemma 8.32, and the fact that the labels belong to $\{-1, 1\}$. Since $n_R = \hat{\tau} n$, we get $|\ell(\boldsymbol{M}, \mathbf{z}, R) - \ell(\boldsymbol{M}^i, \mathbf{z}, R^i)| \leq \frac{1}{\gamma n}(\frac{1}{\lambda \gamma} + \frac{2}{\hat{\tau}})$. □

Bibliography

M. Ehsan Abbasnejad, Dhanesh Ramachandram, and Mandava Rajeswari. A survey of the state of the art in learning the kernels. *Knowledge and Information Systems (KAIS)*, 31(2):193–221, 2012. DOI: 10.1007/s10115-011-0404-6. 2

Noga Alon, Shai Ben-David, Nicolò Cesa-Bianchi, and David Haussler. Scale-sensitive dimensions, uniform convergence, and learnability. *Journal of the ACM*, 44(4):615–631, 1997. DOI: 10.1145/263867.263927. 61

Naomi S. Altman. An introduction to kernel and nearest-neighbor nonparametric regression. *The American Statistician*, 46(3):175–185, 1992. DOI: 10.1080/00031305.1992.10475879. 14

F. Gregory Ashby and Nancy A. Perrin. Toward a unified theory of similarity and recognition. *Psychological Review*, 95(1):124–150, 1988. DOI: 10.1037/0033-295X.95.1.124. 1, 26

Francis R. Bach, Rodolphe Jenatton, Julien Mairal, and Guillaume Obozinski. Optimization with Sparsity-Inducing Penalties. *Foundations and Trends in Machine Learning (FTML)*, 4(1): 1–106, 2012. DOI: 10.1561/2200000015. 23, 25

Ricardo Baeza-Yates and Berthier Ribeiro-Neto. *Modern Information Retrieval.* Addison-Wesley, 1999. 10, 15, 27, 96

Mahdieh S. Baghshah and Saeed B. Shouraki. Semi-Supervised Metric Learning Using Pairwise Constraints. In *Proceedings of the 20th International Joint Conference on Artificial Intelligence (IJCAI)*, pages 1217–1222, 2009. 48

Maria-Florina Balcan and Avrim Blum. On a Theory of Learning with Similarity Functions. In *Proceedings of the 23rd International Conference on Machine Learning (ICML)*, pages 73–80, 2006. DOI: 10.1145/1143844.1143854. 75, 77

Maria-Florina Balcan, Avrim Blum, and Nathan Srebro. Improved Guarantees for Learning via Similarity Functions. In *Proceedings of the 21st Annual Conference on Learning Theory (COLT)*, pages 287–298, 2008a. 57, 63, 75, 76, 77

Maria-Florina Balcan, Avrim Blum, and Santosh Vempala. A Discriminative Framework for Clustering via Similarity Functions. In *ACM Symposium on Theory of Computing (STOC)*, pages 671–680, 2008b. DOI: 10.1145/1374376.1374474. 101

Peter L. Bartlett and Shahar Mendelson. Rademacher and Gaussian Complexities: Risk Bounds and Structural Results. *Journal of Machine Learning Research (JMLR)*, 3:463–482, 2002. 61, 62, 80

Jonathan Baxter and Peter L. Bartlett. The Canonical Distortion Measure in Feature Space and 1-NN Classification. In *Advances in Neural Information Processing Systems (NIPS) 10*, 1997. 1

Mikhail Belkin and Partha Niyogi. Semi-Supervised Learning on Riemannian Manifolds. *Machine Learning Journal (MLJ)*, 56(1–3):209–239, 2004. DOI: 10.1023/B:MACH.0000033120.25363.1e. 47

Aurélien Bellet. *Supervised Metric Learning with Generalization Guarantees*. PhD thesis, University of Saint-Etienne, 2012. 61

Aurélien Bellet and Amaury Habrard. Robustness and Generalization for Metric Learning. *Neurocomputing*, 151(1):259–267, 2015. DOI: 10.1016/j.neucom.2014.09.044. 62, 68, 72

Aurélien Bellet, Amaury Habrard, and Marc Sebban. Learning Good Edit Similarities with Generalization Guarantees. In *Proceedings of the European Conference on Machine Learning and Principles and Practice of Knowledge Discovery in Databases (ECML/PKDD)*, pages 188–203, 2011. DOI: 10.1007/978-3-642-23780-5_22. 57, 63, 99

Aurélien Bellet, Amaury Habrard, and Marc Sebban. Good edit similarity learning by loss minimization. *Machine Learning Journal (MLJ)*, 89(1):5–35, 2012a. DOI: 10.1007/s10994-012-5293-8. 57, 63

Aurélien Bellet, Amaury Habrard, and Marc Sebban. Similarity Learning for Provably Accurate Sparse Linear Classification. In *Proceedings of the 29th International Conference on Machine Learning (ICML)*, pages 1871–1878, 2012b. 27, 63, 74, 75, 77, 102

Aurélien Bellet, Amaury Habrard, and Marc Sebban. A Survey on Metric Learning for Feature Vectors and Structured Data. Technical report, arXiv:1306.6709, 2013. 4

Shai Ben-David, John Blitzer, Koby Crammer, Alex Kulesza, Fernando Pereira, and Jennifer Wortman Vaughan. A theory of learning from different domains. *Machine Learning Journal (MLJ)*, 79(1-2):151–175, 2010. DOI: 10.1007/s10994-009-5152-4. 48

Yoshua Bengio. Learning deep architectures for AI. *Foundations and Trends in Machine Learning (FTML)*, 2(1):1–127, 2009. DOI: 10.1561/2200000006. 35

Marc Bernard, Amaury Habrard, and Marc Sebban. Learning Stochastic Tree Edit Distance. In *Proceedings of the 17th European Conference on Machine Learning (ECML)*, pages 42–53, 2006. DOI: 10.1007/11871842_9. 58

Marc Bernard, Laurent Boyer, Amaury Habrard, and Marc Sebban. Learning probabilistic models of tree edit distance. *Pattern Recognition (PR)*, 41(8):2611–2629, 2008. DOI: 10.1016/j.patcog.2008.01.011. 58

Enrico Bertini, Andrada Tatu, and Daniel Keim. Quality Metrics in High-Dimensional Data Visualization: An Overview and Systematization. *IEEE Transactions on Visualization and Computer Graphics (TVCG)*, 17(12):2203–2212, 2011. DOI: 10.1109/TVCG.2011.229. 15

Jinbo Bi, Dijia Wu, Le Lu, Meizhu Liu, Yimo Tao, and Matthias Wolf. AdaBoost on low-rank PSD matrices for metric learning. In *Proceedings of the IEEE Conference on Computer Vision and Pattern Recognition (CVPR)*, pages 2617–2624, 2011. DOI: 10.1109/CVPR.2011.5995363. 26

Wei Bian and Dacheng Tao. Learning a Distance Metric by Empirical Loss Minimization. In *Proceedings of the 22nd International Joint Conference on Artificial Intelligence (IJCAI)*, pages 1186–1191, 2011. DOI: 10.5591/978-1-57735-516-8/IJCAI11-202. 62

Wei Bian and Dacheng Tao. Constrained Empirical Risk Minimization Framework for Distance Metric Learning. *IEEE Transactions on Neural Networks and Learning Systems (TNNLS)*, 23 (8):1194–1205, 2012. DOI: 10.1109/TNNLS.2012.2198075. 62

Mikhail Bilenko and Raymond J. Mooney. Adaptive Duplicate Detection Using Learnable String Similarity Measures. In *Proceedings of the 9th ACM SIGKDD International Conference on Knowledge Discovery and Data Mining*, pages 39–48, 2003. DOI: 10.1145/956750.956759. 55

Mikhail Bilenko, Sugato Basu, and Raymond J. Mooney. Integrating Constraints and Metric Learning in Semi-Supervised Clustering. In *Proceedings of the 21st International Conference on Machine Learning (ICML)*, pages 81–88, 2004. DOI: 10.1145/1015330.1015360. 47

Philip Bille. A survey on tree edit distance and related problems. *Theoretical Computer Science (TCS)*, 337(1-3):217–239, 2005. DOI: 10.1016/j.tcs.2004.12.030. 13

Oren Boiman, Eli Shechtman, and Michal Irani. In defense of nearest-neighbor based image classification. In *Proceedings of the IEEE Conference on Computer Vision and Pattern Recognition (CVPR)*, 2008. DOI: 10.1109/CVPR.2008.4587598. 89

Léon Bottou and Olivier Bousquet. The Tradeoffs of Large Scale Learning. In *Advances in Neural Information Processing Systems (NIPS) 20*, pages 161–168, 2007. 28, 31

Olivier Bousquet and André Elisseeff. Stability and Generalization. *Journal of Machine Learning Research (JMLR)*, 2:499–526, 2002. DOI: 10.1162/153244302760200704. 62, 64, 65, 78, 103

Stephen Boyd and Lieven Vandenberghe. *Convex Optimization*. Cambridge University Press, 2004. DOI: 10.1017/CBO9780511804441. 3

Stephen P. Boyd, Neal Parikh, Eric Chu, Borja Peleato, and Jonathan Eckstein. Distributed Optimization and Statistical Learning via the Alternating Direction Method of Multipliers. *Foundations and Trends in Machine Learning (FTML)*, 3(1):1–122, 2011. DOI: 10.1561/2200000016. 30

Laurent Boyer, Amaury Habrard, and Marc Sebban. Learning Metrics between Tree Structured Data: Application to Image Recognition. In *Proceedings of the 18th European Conference on Machine Learning (ECML)*, pages 54–66, 2007. DOI: 10.1007/978-3-540-74958-5_9. 58

Laurent Boyer, Yann Esposito, Amaury Habrard, José Oncina, and Marc Sebban. SEDiL: Software for Edit Distance Learning. In *Proceedings of the European Conference on Machine Learning and Principles and Practice of Knowledge Discovery in Databases (ECML/PKDD)*, pages 672–677, 2008. URL `http://labh-curien.univ-st-etienne.fr/SEDiL/`. DOI: 10.1007/978-3-540-87481-2_45. 54

Laurent Boyer, Olivier Gandrillon, Amaury Habrard, Mathilde Pellerin, and Marc Sebban. Learning constrained edit state machines. In *Proceedings of the 21st IEEE International Conference on Tools with Artificial Intelligence (ICTAI)*, pages 734–741, 2009. DOI: 10.1109/ICTAI.2009.27. 93

Lev M. Bregman. The relaxation method of finding the common points of convex sets and its application to the solution of problems in convex programming. *USSR Computational Mathematics and Mathematical Physics*, 7(3):200–217, 1967. DOI: 10.1016/0041-5553(67)90040-7. 40

Leo Breiman. Random Forests. *Machine Learning Journal (MLJ)*, 45(1):5–32, 2001. DOI: 10.1023/A:1010933404324. 41

Bin Cao, Xiaochuan Ni, Jian-Tao Sun, Gang Wang, and Qiang Yang. Distance Metric Learning under Covariate Shift. In *Proceedings of the 22nd International Joint Conference on Artificial Intelligence (IJCAI)*, pages 1204–1210, 2011. DOI: 10.5591/978-1-57735-516-8/IJCAI11-205. 49

Qiong Cao, Zheng-Chu Guo, and Yiming Ying. Generalization Bounds for Metric and Similarity Learning. Technical report, University of Exeter, July 2012a. arXiv:1207.5437. 62, 63, 80

Qiong Cao, Yiming Ying, and Peng Li. Distance Metric Learning Revisited. In *Proceedings of the European Conference on Machine Learning and Principles and Practice of Knowledge Discovery in Databases (ECML/PKDD)*, pages 283–298, 2012b. DOI: 10.1007/978-3-642-33460-3_24. 20

Rich Caruana. Multitask Learning. *Machine Learning Journal (MLJ)*, 28(1):41–75, 1997. DOI: 10.1023/A:1007379606734. 43

Nicoló Cesa-Bianchi and Claudio Gentile. Improved Risk Tail Bounds for On-Line Algorithms. *IEEE Transactions on Information Theory (TIT)*, 54(1):386–390, 2008. DOI: 10.1109/TIT.2007.911292. 62

Nicolò Cesa-Bianchi and Gabor Lugosi. *Prediction, Learning, and Games*. Cambridge University Press, 2006. DOI: 10.1017/CBO9780511546921. 28

Ratthachat Chatpatanasiri, Teesid Korsrilabutr, Pasakorn Tangchanachaianan, and Boonserm Kijsirikul. A new kernelization framework for Mahalanobis distance learning algorithms. *Neurocomputing*, 73:1570–1579, 2010. DOI: 10.1016/j.neucom.2009.11.037. 34

Gal Chechik, Uri Shalit, Varun Sharma, and Samy Bengio. An Online Algorithm for Large Scale Image Similarity Learning. In *Advances in Neural Information Processing Systems (NIPS) 22*, pages 306–314, 2009. 27, 29, 74

Gal Chechik, Varun Sharma, Uri Shalit, and Samy Bengio. Large Scale Online Learning of Image Similarity Through Ranking. *Journal of Machine Learning Research (JMLR)*, 11:1109–1135, 2010. DOI: 10.1145/1756006.1756042. 27, 29

Minmin Chen, Zhixiang Eddie Xu, Kilian Q. Weinberger, and Fei Sha. Marginalized Denoising Autoencoders for Domain Adaptation. In *Proceedings of the 29th International Conference on Machine Learning (ICML)*, 2012. 101

Li Cheng. Riemannian Similarity Learning. In *Proceedings of the 30th International Conference on Machine Learning (ICML)*, 2013. 27

Sumit Chopra, Raia Hadsell, and Yann LeCun. Learning a Similarity Metric Discriminatively, with Application to Face Verification. In *Proceedings of the IEEE Conference on Computer Vision and Pattern Recognition (CVPR)*, pages 539–546, 2005. DOI: 10.1109/CVPR.2005.202. 35

Ramazan Gokberk Cinbis, Jakob J. Verbeek, and Cordelia Schmid. Unsupervised metric learning for face identification in tv video. In *Proceedings of the IEEE International Conference on Computer Vision (ICCV)*, pages 1559–1566, 2011. DOI: 10.1109/ICCV.2011.6126415. 87

Stéphan Clémençon, Aurélien Bellet, and Igor Colin. Scaling-up Empirical Risk Minimization: Optimization of Incomplete U-statistics. Technical report, arXiv:1501.02629, 2015. DOI: 10.1109/ICCV.2011.6126415. 28, 31

Corinna Cortes and Vladimir Vapnik. Support-Vector Networks. *Machine Learning Journal (MLJ)*, 20(3):273–297, 1995. DOI: 10.1023/A:1022627411411. 14, 23

Corinna Cortes, Patrick Haffner, and Mehryar Mohri. Rational Kernels: Theory and Algorithms. *Journal of Machine Learning Research (JMLR)*, 5:1035–1062, 2004. 14

Thomas Cover and Peter Hart. Nearest neighbor pattern classification. *IEEE Transactions on Information Theory (TIT)*, 13(1):21–27, 1967. DOI: 10.1109/TIT.1967.1053964. 1, 14

Koby Crammer and Gal Chechik. Adaptive Regularization for Weight Matrices. In *Proceedings of the 29th International Conference on Machine Learning (ICML)*, 2012. 30

Koby Crammer, Ofer Dekel, Joseph Keshet, Shai Shalev-Shwartz, and Yoram Singer. Online Passive-Aggressive Algorithms. *Journal of Machine Learning Research (JMLR)*, 7:551–585, 2006. 29

Marco Cuturi and David Avis. Ground Metric Learning. Technical report, Kyoto University, 2011. 1110.2306. 51

Nilesh N. Dalvi, Philip Bohannon, and Fei Sha. Robust web extraction: an approach based on a probabilistic tree-edit model. In *Proceedings of the ACM SIGMOD International Conference on Management of data (COMAD)*, pages 335–348, 2009. DOI: 10.1145/1559845.1559882. 58

Jason V. Davis and Inderjit S. Dhillon. Structured metric learning for high dimensional problems. In *Proceedings of the 14th ACM SIGKDD International Conference on Knowledge Discovery and Data Mining*, pages 195–203, 2008. DOI: 10.1145/1401890.1401918. 96

Jason V. Davis, Brian Kulis, Prateek Jain, Suvrit Sra, and Inderjit S. Dhillon. Information-theoretic metric learning. In *Proceedings of the 24th International Conference on Machine Learning (ICML)*, pages 209–216, 2007. DOI: 10.1145/1273496.1273523. 24, 34

Margaret O. Dayhoff, Robert M. Schwartz, and Bruce C. Orcutt. A model of evolutionary change in proteins. *Atlas of protein sequence and structure*, 5(3):345–351, 1978. 54

Ofer Dekel, Ran Gilad-Bachrach, Ohad Shamir, and Lin Xiao. Optimal Distributed Online Prediction Using Mini-Batches. *Journal of Machine Learning Research (JMLR)*, 13:165–202, 2012. 30

Arthur P. Dempster, Nan M. Laird, and Donald B. Rubin. Maximum likelihood from incomplete data via the EM algorithm. *Journal of the Royal Statistical Society, Series B*, 39(1):1–38, 1977. 54

J. Deng, W. Dong, R. Socher, L.J. Li, K. Li, and L. Fei-Fei. Imagenet: A large-scale hierarchical image database. In *Proceedings of the IEEE Conference on Computer Vision and Pattern Recognition (CVPR)*, 2009. DOI: 10.1109/CVPR.2009.5206848. 88

Jia Deng, Alexander C. Berg, and Li Fei-Fei. Hierarchical semantic indexing for large scale image retrieval. In *Proceedings of the IEEE Conference on Computer Vision and Pattern Recognition (CVPR)*, pages 785–792, 2011. DOI: 10.1109/CVPR.2011.5995516. 10

Li Deng and Dong Yu. Deep Learning: Methods and Applications. *Foundations and Trends in Signal Processing (FTSP)*, 7(3-4):197–387, 2014. DOI: 10.1561/2000000039. 35

Inderjit S. Dhillon and Joel A. Tropp. Matrix Nearness Problems with Bregman Divergences. *SIAM Journal on Matrix Analysis and Applications*, 29(4):1120–1146, 2007. DOI: 10.1137/060649021. 45

Huyen Do, Alexandros Kalousis, Jun Wang, and Adam Woznica. A metric learning perspective of SVM: on the relation of LMNN and SVM. In *Proceedings of the 15th International Conference on Artificial Intelligence and Statistics (AISTATS)*, pages 308–317, 2012. 22

John Duchi, Elad Hazan, and Yoram Singer. Adaptive Subgradient Methods for Online Learning and Stochastic Optimization. *Journal of Machine Learning Research (JMLR)*, 12:2121–2159, 2011. 30

John C. Duchi, Alekh Agarwal, and Martin J. Wainwright. Dual Averaging for Distributed Optimization: Convergence Analysis and Network Scaling. *IEEE Transactions on Automatic Control (TAC)*, 57(3):592–606, 2012. DOI: 10.1109/TAC.2011.2161027. 30

Martin Emms. On Stochastic Tree Distances and Their Training via Expectation-Maximisation. In *Proceedings of the 1st International Conference on Pattern Recognition Applications and Methods (ICPRAM)*, pages 144–153, 2012. 58

Theodoros Evgeniou and Massimiliano Pontil. Regularized multi-task learning. In *Proceedings of the 10th ACM SIGKDD International Conference on Knowledge Discovery and Data Mining*, pages 109–117, 2004. DOI: 10.1145/1014052.1014067. 43

R. Fang, K. Tang, N. Snavely, and T. Chen. Towards computational models of kinship verification. In *Proceedings of the International Conference on Image Processing (ICIP)*, page 1577–1580, 2010. DOI: 10.1109/ICIP.2010.5652590. 88

Bao Q. Feng. Equivalence constants for certain matrix norms. *Linear Algebra and Its Applications*, 374:247–253, 2003. DOI: 10.1016/S0024-3795(03)00616-5. 110

Imola K. Fodor. A Survey of Dimension Reduction Techniques. Technical report, Lawrence Livermore National Laboratory, 2002. UCRL-ID- 148494. DOI: 10.2172/15002155. 3

Cedric Frambourg, Ahlame Douzal Chouakria, and Eric Gaussier. Learning multiple temporal matching for time series classification. In *Proceedings of the 12th International Symposium on Advances in Intelligent Data Analysis (IDA)*, volume 8207, pages 198–209, 2013. DOI: 10.1007/978-3-642-41398-8_18. 59

Yoav Freund and Robert E. Schapire. A Decision-Theoretic Generalization of On-Line Learning and an Application to Boosting. In *Proceedings of the 2nd European Conference on Computational Learning Theory (EuroCOLT)*, pages 23–37, 1995. DOI: 10.1006/jcss.1997.1504. 25

Jerome H. Friedman. Flexible Metric Nearest Neighbor Classification. Technical report, Department of Statistics, Stanford University, 1994. DOI: 10.1109/IGARSS.2008.4779695. 1

Jerome H. Friedman. Greedy Function Approximation: A Gradient Boosting Machine. *Annals of Statistics (AOS)*, 29(5):1189–1232, 2001. DOI: 10.1214/aos/1013203451. 36

Andrea Frome, Yoram Singer, Fei Sha, and Jitendra Malik. Learning Globally-Consistent Local Distance Functions for Shape-Based Image Retrieval and Classification. In *Proceedings of the IEEE International Conference on Computer Vision (ICCV)*, pages 1–8, 2007. DOI: 10.1109/ICCV.2007.4408839. 91

Keinosuke Fukunaga. *Introduction to Statistical Pattern Recognition.* Academic Press, 1990. 1

Xinbo Gao, Bing Xiao, Dacheng Tao, and Xuelong Li. A survey of graph edit distance. *Pattern Analysis and Applications (PAA)*, 13(1):113–129, 2010. DOI: 10.1007/s10044-008-0141-y. 13

Xingyu Gao, Steven C.H. Hoi, Yongdong Zhang, Ji Wan, and Jintao Li. SOML: Sparse Online Metric Learning with Application to Image Retrieval. In *Proceedings of the 28th AAAI Conference on Artificial Intelligence*, pages 1206–1212, 2014. 32

Thomas Gärtner. A survey of kernels for structured data. *SIGKDD Explorations*, 5(1):49–58, 2003. DOI: 10.1145/959242.959248. 15

R. Gaudin and Nicolas Nicoloyannis. An adaptable time warping distance for time series learning. In *Proceedings of the 5th International Conference on Machine Learning and Applications (ICMLA)*, December 2006. DOI: 10.1109/ICMLA.2006.12. 59

Bo Geng, Dacheng Tao, and Chao Xu. DAML: Domain Adaptation Metric Learning. *IEEE Transactions on Image Processing (TIP)*, 20(10):2980–2989, 2011. DOI: 10.1109/TIP.2011.2134107. 49

Amir Globerson and Sam T. Roweis. Metric Learning by Collapsing Classes. In *Advances in Neural Information Processing Systems (NIPS) 18*, pages 451–458, 2005. 21

Jacob Goldberger, Sam Roweis, Geoff Hinton, and Ruslan Salakhutdinov. Neighbourhood Components Analysis. In *Advances in Neural Information Processing Systems (NIPS) 17*, pages 513–520, 2004. 20

Robert L. Goldstone, Douglas L. Medin, and Jamin Halberstadt. Similarity in context. *Memory & Cognition*, 25(2):237–255, 1997. DOI: 10.3758/BF03201115. 1

Mehmet Gönen and Ethem Alpaydin. Multiple Kernel Learning Algorithms. *Journal of Machine Learning Research (JMLR)*, 12:2211–2268, 2011. 2

Arthur Gretton, Karsten M. Borgwardt, Malte J. Rasch, Bernhard Schölkopf, and Alexander J. Smola. A Kernel Method for the Two-Sample-Problem. In *Advances in Neural Information Processing Systems (NIPS) 19*, pages 513–520, 2006. 49

Matthieu Guillaumin, Thomas Mensink, Jakob Verbeek, and Cordelia Schmid. Tagprop: Discriminative metric learning in nearest neighbor models for image auto-annotation. In *Proceedings of the IEEE International Conference on Computer Vision (ICCV)*, pages 309–316, 2009a. DOI: 10.1109/ICCV.2009.5459266. 90

Matthieu Guillaumin, Jakob J. Verbeek, and Cordelia Schmid. Is that you? Metric learning approaches for face identification. In *Proceddings of the IEEE International Conference on Computer Vision (ICCV)*, pages 498–505, 2009b. DOI: 10.1109/ICCV.2009.5459197. 87

Matthieu Guillaumin, Jakob Verbeek, and Cordelia Schmid. Multiple instance metric learning from automatically labeled bags of faces. In *Proceedings of the 10th European Conference on Computer Vision (ECCV)*, volume 6311, pages 634–647, 2010. DOI: 10.1007/978-3-642-15549-9_46. 87

Zheng-Chu Guo and Yiming Ying. Guaranteed Classification via Regularized Similarity Learning. *Neural Computation*, 26(3):497–522, 2014. DOI: 10.1162/NECO_a_00556. 63, 80, 81, 82, 102

James L. Hafner, Harpreet S. Sawhney, William Equitz, Myron Flickner, and Wayne Niblack. Efficient Color Histogram Indexing for Quadratic Form Distance Functions. *IEEE Transactions on Pattern Analysis and Machine Intelligence (TPAMI)*, 17(7):729–736, 1995. DOI: 10.1109/34.391417. 11

Ulrike Hahn, Nick Chater, and Lucy B. Richardson. Similarity as transformation. *Cognition*, 87 (1):1–32, 2003. DOI: 10.1016/S0010-0277(02)00184-1. 1

Trevor Hastie and Robert Tibshirani. Discriminant Adaptive Nearest Neighbor Classification. *IEEE Transactions on Pattern Analysis and Machine Intelligence (TPAMI)*, 18(6):607–616, 1996. DOI: 10.1109/34.506411. 1

Søren Hauberg, Oren Freifeld, and Michael J. Black. A Geometric take on Metric Learning. In *Advances in Neural Information Processing Systems (NIPS) 25*, pages 2033–2041, 2012. 38

Yujie He, Wenlin Chen, and Yixin Chen. Kernel Density Metric Learning. Technical report, Washington University in St. Louis, 2013. 34

Steven Henikoff and Jorja G. Henikoff. Amino acid substitution matrices from protein blocks. *Proceedings of the National Academy of Sciences of the United States of America*, 89(22):10915–10919, 1992. DOI: 10.1073/pnas.89.22.10915. 54

Steven C. Hoi, Wei Liu, Michael R. Lyu, and Wei-Ying Ma. Learning Distance Metrics with Contextual Constraints for Image Retrieval. In *Proceedings of the IEEE Conference on Computer Vision and Pattern Recognition (CVPR)*, pages 2072–2078, 2006. DOI: 10.1109/CVPR.2006.167. 34

Steven C. Hoi, Wei Liu, and Shih-Fu Chang. Semi-supervised distance metric learning for Collaborative Image Retrieval. In *Proceedings of the IEEE Conference on Computer Vision and Pattern Recognition (CVPR)*, 2008. DOI: 10.1109/CVPR.2008.4587351. 47, 48

Steven C. Hoi, Wei Liu, and Shih-Fu Chang. Semi-supervised distance metric learning for collaborative image retrieval and clustering. *ACM Transactions on Multimedia Computing, Communications, and Applications (TOMCCAP)*, 6(3), 2010. DOI: 10.1145/1823746.1823752. 47, 48

Junlin Hu, Jiwen Lu, and Yap-Peng Tan. Discriminative Deep Metric Learning for Face Verification in the Wild. In *IEEE Conference on Computer Vision and Pattern Recognition (CVPR)*, pages 1875–1882, 2014. DOI: 10.1145/1823746.1823752. 35

G. Huang, M. Ramesh, T. Berg, and E. Learned-Miller. Labeled faces in the wild: a database for studying face recognition in unconstrained environments. Technical Report 07-49, University of Massachusetts, Amherst, 2007. 86

Prateek Jain, Brian Kulis, Inderjit S. Dhillon, and Kristen Grauman. Online Metric Learning and Fast Similarity Search. In *Advances in Neural Information Processing Systems (NIPS) 21*, pages 761–768, 2008. 24, 29

Prateek Jain, Brian Kulis, and Inderjit S. Dhillon. Inductive Regularized Learning of Kernel Functions. In *Advances in Neural Information Processing Systems (NIPS) 23*, pages 946–954, 2010. 34

Prateek Jain, Brian Kulis, Jason V. Davis, and Inderjit S. Dhillon. Metric and Kernel Learning Using a Linear Transformation. *Journal of Machine Learning Research (JMLR)*, 13:519–547, 2012. 34

Young-Seon Jeong, Myong K. Jeong, and Olufemi A. Omitaomu. Weighted dynamic time warping for time series classification. *Pattern Recognition*, 44(9):2231–2240, 2010. DOI: 10.1016/j.patcog.2010.09.022. 59

Rong Jin, Shijun Wang, and Yang Zhou. Regularized Distance Metric Learning: Theory and Algorithm. In *Advances in Neural Information Processing Systems (NIPS) 22*, pages 862–870, 2009. 29, 62, 64

Thorsten Joachims, Thomas Finley, and Chun-Nam J. Yu. Cutting-plane training of structural SVMs. *Machine Learning Journal (MLJ)*, 77(1):27–59, 2009. DOI: 10.1007/s10994-009-5108-8. 46

Purushottam Kar and Prateek Jain. Similarity-based Learning via Data Driven Embeddings. In *Advances in Neural Information Processing Systems (NIPS) 24*, 2011. 77

Purushottam Kar, Bharath Sriperumbudur Prateek Jain, and Harish Karnick. On the Generalization Ability of Online Learning Algorithms for Pairwise Loss Functions. In *Proceedings of the 30th International Conference on Machine Learning*, 2013. 63

Tsuyoshi Kato and Nozomi Nagano. Metric learning for enzyme active-site search. *Bioinformatics*, 26(21):2698–2704, 2010. DOI: 10.1093/bioinformatics/btq519. 94

Leonard Kaufman and Peter Rousseeuw. Clustering by Means of Medoids. In Y. Dodge, Elsevier/North Holland, editor, *Statistical Data adlysis based on the L1 norm and Related Methods*, pages 405–416, 1987. 15

Dor Kedem, Stephen Tyree, Kilian Weinberger, Fei Sha, and Gert Lanckriet. Non-linear Metric Learning. In *Advances in Neural Information Processing Systems (NIPS) 25*, pages 2582–2590, 2012. 35, 50, 101

Anne-Louise Klaus and Chi-Kwong Li. Isometries for the vector (p,q) norm and the induced (p,q) norm. *Linear & Multilinear Algebra*, 38(4):315–332, 1995. DOI: 10.1080/03081089508818368. 110

Andrei N. Kolmogorov and Vassili M. Tikhomirov. ϵ-entropy and ϵ-capacity of sets in functional spaces. *American Mathematical Society Translations*, 2(17):277–364, 1961. 69

Vladimir Koltchinskii. Rademacher penalties and structural risk minimization. *IEEE Transactions on Information Theory (TIT)*, 47(5):1902–1914, 2001. DOI: 10.1109/18.930926. 61

J. B. Kruskall and M. Liberman. The symmetric time warping algorithm: From continuous to discrete. *Time warps, string edits and macromolecules*, 1983. 58

Brian Kulis, Mátyás A. Sustik, and Inderjit S. Dhillon. Learning low-rank kernel matrices. In *Proceedings of the 23rd International Conference on Machine Learning (ICML)*, pages 505–512, 2006. DOI: 10.1145/1143844.1143908. 24

Brian Kulis, Prateek Jain, and Kristen Grauman. Fast Similarity Search for Learned Metrics. *IEEE Transactions on Pattern Analysis and Machine Intelligence (TPAMI)*, 31(12):2143–2157, 2009. DOI: 10.1109/TPAMI.2009.151. 24

Brian Kulis, Kate Saenko, and Trevor Darrell. What you saw is not what you get: Domain adaptation using asymmetric kernel transforms. In *Proceedings of the IEEE Conference on Computer Vision and Pattern Recognition (CVPR)*, pages 1785–1792, 2011. DOI: 10.1109/CVPR.2011.5995702. 48

Solomon Kullback and Richard Leibler. On Information and Sufficiency. *Annals of Mathematical Statistics*, 22(1):79–86, 1951. DOI: 10.1214/aoms/1177729694. 41

M. Pawan Kumar, Philip H. S. Torr, and Andrew Zisserman. An Invariant Large Margin Nearest Neighbour Classifier. In *Proceedings of the IEEE International Conference on Computer Vision (ICCV)*, pages 1–8, 2007. DOI: 10.1109/ICCV.2007.4409041. 101

Gautam Kunapuli and Jude Shavlik. Mirror Descent for Metric Learning: A Unified Approach. In *Proceedings of the European Conference on Machine Learning and Principles and Practice of Knowledge Discovery in Database (ECML/PKDD)*, pages 859–874, 2012. DOI: 10.1007/978-3-642-33460-3_60. 25

Rémi Lajugie, Sylvain Arlot, and Francis Bach. Large-Margin Metric Learning for Constrained Partitioning Problems. In *Proceedings of the 31st International Conference on Machine Learning (ICML)*, 2014a. 101

Rémi Lajugie, Damien Garreau, Francis R. Bach, and Sylvain Arlot. Metric learning for temporal sequence alignment. In *Advances in Neural Information Processing Systems (NIPS) 27*, pages 1817–1825, 2014b. 59, 99

Nicolas Le Roux, Mark W. Schmidt, and Francis Bach. A Stochastic Gradient Method with an Exponential Convergence Rate for Finite Training Sets. In *Advances in Neural Information Processing Systems (NIPS) 25*, pages 2672–2680, 2012. 30

Guy Lebanon. Metric Learning for Text Documents. *IEEE Transactions on Pattern Analysis and Machine Intelligence (TPAMI)*, 28(4):497–508, 2006. DOI: 10.1109/TPAMI.2006.77. 96

Jung-Eun Lee, Rong Jin, and Anil K. Jain. Rank-based distance metric learning: An application to image retrieval. In *Proceedings of the IEEE Conference on Computer Vision and Pattern Recognition (CVPR)*, 2008. DOI: 10.1109/CVPR.2008.4587389. 95

Christina S. Leslie, Eleazar Eskin, and William S. Noble. The Spectrum Kernel: A String Kernel for SVM Protein Classification. In *Pacific Symposium on Biocomputing*, pages 566–575, 2002a. 13

Christina S. Leslie, Eleazar Eskin, Jason Weston, and William S. Noble. Mismatch String Kernels for SVM Protein Classification. In *Advances in Neural Information Processing Systems (NIPS) 15*, pages 1417–1424, 2002b. 14

Vladimir I. Levenshtein. Binary codes capable of correcting deletions, insertions and reversals. *Soviet Physics-Doklandy*, 6:707–710, 1966. 12

Fei-Fei Li and Pietro Perona. A Bayesian Hierarchical Model for Learning Natural Scene Categories. In *Proceedings of the IEEE Conference on Computer Vision and Pattern Recognition*, pages 524–531, 2005. DOI: 10.1109/CVPR.2005.16. 11, 53, 85

Haifeng Li and Tao Jiang. A class of edit kernels for SVMs to predict translation initiation sites in eukaryotic mRNAs. In *Proceedings of the 8th Annual International Conference on Research in Computational Molecular Biology (RECOMB)*, pages 262–271, 2004. DOI: 10.1145/974614.974649. 14

Xi Li, Chunhua Shen, Qinfeng Shi, Anthony R. Dick, and Anton van den Hengel. Non-sparse linear representations for visual tracking with online reservoir metric learning. In *Proceedings of the IEEE Conference on Computer Vision and Pattern Recognition (CVPR)*, 2012. DOI: 10.1109/CVPR.2012.6247872. 92

Daryl Lim and Gert R. Lanckriet. Efficient Learning of Mahalanobis Metrics for Ranking. In *Proceedings of the 31st International Conference on Machine Learning (ICML)*, pages 1980–1988, 2014. 47

Daryl K. Lim, Brian McFee, and Gert Lanckriet. Robust Structural Metric Learning. In *Proceedings of the 30th International Conference on Machine Learning (ICML)*, 2013. 47

Nick Littlestone. Learning Quickly When Irrelevant Attributes Abound: A New Linear-Threshold Algorithm. *Machine Learning Journal (MLJ)*, 2(4):285–318, 1988. DOI: 10.1023/A:1022869011914. 28

Kuan Liu, Aurélien Bellet, and Fei Sha. Similarity Learning for High-Dimensional Sparse Data. Technical report, arXiv:1411.2374, 2014. 32

Meizhu Liu and Baba C. Vemuri. A Robust and Efficient Doubly Regularized Metric Learning Approach. In *Proceedings of the 12th European Conference on Computer Vision (ECCV)*, pages 646–659, 2012. DOI: 10.1007/978-3-642-33765-9_46. 26

Tie-Yan Liu. Learning to Rank for Information Retrieval. *Foundations and Trends in Information Retrieval (FTIR)*, 3(3):225–331, 2009. DOI: 10.1561/1500000016. 46

Wei Liu, Shiqian Ma, Dacheng Tao, Jianzhuang Liu, and Peng Liu. Semi-Supervised Sparse Metric Learning using Alternating Linearization Optimization. In *Proceedings of the 16th ACM SIGKDD International Conference on Knowledge Discovery and Data Mining*, pages 1139–1148, 2010. DOI: 10.1145/1835804.1835947. 48

Stuart P. Lloyd. Least squares quantization in PCM. *IEEE Transactions on Information Theory (TIT)*, 28:129–137, 1982. DOI: 10.1109/TIT.1982.1056489. 1, 15

Huma Lodhi, Craig Saunders, John Shawe-Taylor, Nello Cristianini, and Chris Watkins. Text Classification using String Kernels. *Journal of Machine Learning Research (JMLR)*, 2:419–444, 2002. DOI: 10.1162/153244302760200687. 14, 53

Jiwen Lu, Junlin Hu, Xiuzhuang Zhou, Yuanyuan Shang, Yap-Peng Tan, and Gang Wang. Neighborhood repulsed metric learning for kinship verification. In *Proceedings of the IEEE Conference on Computer Vision and Pattern Recognition (CVPR)*, pages 2594–2601, 2012a. DOI: 10.1109/CVPR.2012.6247978. 88

Jiwen Lu, Gang Wang, and Thomas S. Huang. Gait-based gender classification in unconstrained environments. In *Proceedings of the 21st International Conference on Pattern Recognition (ICPR)*, pages 3284–3287, 2012b. DOI: 10.1109/ITNG.2013.134. 91

Prasanta Chandra Mahalanobis. On the generalised distance in statistics. *Proceedings of the National Institute of Sciences of India*, 2(1):49–55, 1936. 9

Christopher D. Manning, Prabhakar Raghavan, and Hinrich Schütze. *Introduction to Information Retrieval*. Cambridge University Press, 2008. DOI: 10.1017/CBO9780511809071. 1, 15

Yishay Mansour, Mehryar Mohri, and Afshin Rostamizadeh. Domain Adaptation: Learning Bounds and Algorithms. In *Proceedings of the 22nd Annual Conference on Learning Theory (COLT)*, 2009. 48

Arthur B. Markman and Dedre Gentner. Structural Alignment during Similarity Comparisons. *Cognitive Psychology*, 25(4):431–467, 1993. DOI: 10.1006/cogp.1993.1011. 1

Andrew McCallum, Kedar Bellare, and Fernando Pereira. A Conditional Random Field for Discriminatively-trained Finite-state String Edit Distance. In *Proceedings of the 21st Conference in Uncertainty in Artificial Intelligence (UAI)*, pages 388–395, 2005. 56

Colin McDiarmid. *Surveys in Combinatorics*, chapter On the method of bounded differences, pages 148–188. Cambridge University Press, 1989. 66

Brian McFee and Gert R. G. Lanckriet. Metric Learning to Rank. In *Proceedings of the 27th International Conference on Machine Learning (ICML)*, pages 775–782, 2010. 46, 68, 73

Douglas L. Medin, Robert L. Goldstone, and Dedre Gentner. Respects for similarity. *Psychological Review*, 100(2):254–278, 1993. DOI: 10.1037/0033-295X.100.2.254. 1

Thomas Mensink, Jakob J. Verbeek, Florent Perronnin, and Gabriela Csurka. Metric learning for large scale image classification: Generalizing to new classes at near-zero cost. In *Proceedings of the 10th European Conference on Computer Vision (ECCV)*, pages 488–501, 2012. DOI: 10.1007/978-3-642-33709-3_35. 90

Luisa Micó and Jose Oncina. Comparison of fast nearest neighbour classifiers for handwritten character recognition. *Pattern Recognition Letters (PRL)*, 19:351–356, 1998. DOI: 10.1016/S0167-8655(98)00007-5. 54

Kenta Mikawa, Takashi Ishida, and Masayuki Goto. A proposal of extended cosine measure for distance metric learning in text classification. In *Proceedings of the IEEE International Conference on Systems, Man, and Cybernetics (SMC)*, pages 1741–1746, 2011. DOI: 10.1109/ICSMC.2011.6083923. 96

Kenta Mikawa, Takashi Ishida, and Masayuki Goto. An optimal weighting method by using the category information in text classification based on metric learning. *Industrial Engineering & Management Systems*, 11(1):87–93, 2012. DOI: 10.7232/iems.2012.11.1.087. 96

David W. Mount. *Bioinformatics: Sequence and Genome Analysis*. Cold Spring Harbor Laboratory Press, 2nd edition, 2004. 12

Saul B. Needleman and Christian D. Wunsch. A general method applicable to the search for similarities in the amino acid sequence of two proteins. *Journal of Molecular Biology (JMB)*, 48 (3):443–453, 1970. DOI: 10.1016/0022-2836(70)90057-4. 13

Yurii Nesterov. Smooth minimization of non-smooth functions. *Mathematical Programming*, 103:127–152, 2005. DOI: 10.1007/s10107-004-0552-5. 25

Michel Neuhaus and Horst Bunke. Automatic learning of cost functions for graph edit distance. *Journal of Information Science (JIS)*, 177(1):239–247, 2007. DOI: 10.1016/j.ins.2006.02.013. 58

Behnam Neyshabur, Nati Srebro, Ruslan Salakhutdinov, Yury Makarychev, and Payman Yadollahpour. The Power of Asymmetry in Binary Hashing. In *Advances in Neural Information Processing Systems (NIPS) 26*, pages 2823–2831, 2013. 37

Hieu V. Nguyen and Li Bai. Cosine Similarity Metric Learning for Face Verification. In *Proceedings of the 10th Asian Conference on Computer Vision (ACCV)*, pages 709–720, 2010. DOI: 10.1007/978-3-642-19309-5_55. 27

Nam Nguyen and Yunsong Guo. Metric Learning: A Support Vector Approach. In *Proceedings of the European Conference on Machine Learning and Principles and Practice of Knowledge Discovery in Databases (ECML/PKDD)*, pages 125–136, 2008. DOI: 10.1007/978-3-540-87481-2_9. 22

Mohammad Norouzi, David J. Fleet, and Ruslan Salakhutdinov. Hamming Distance Metric Learning. In *Advances in Neural Information Processing Systems (NIPS) 25*, pages 1070–1078, 2012a. 36

Mohammad Norouzi, Ali Punjani, and David J. Fleet. Fast Search in Hamming Space with Multi-Index Hashing. In *Proceedings of the IEEE Conference on Computer Vision and Pattern Recognition (CVPR)*, 2012b. DOI: 10.1109/CVPR.2012.6248043. 36

Robert M. Nosofsky. Attention, similarity, and the identification-categorization relationship. *Journal of Experimental Psychology: General*, 115(1):39–57, 1986. DOI: 10.1037/0096-3445.115.1.39. 1

Eric Nowak and Frédéric Jurie. Learning visual similarity measures for comparing never seen objects. In *Proceedings of the IEEE Conference on Computer Vision and Pattern Recognition (CVPR)*, 2007. DOI: 10.1109/CVPR.2007.382969. 91

Jose Oncina and Marc Sebban. Learning Stochastic Edit Distance: application in handwritten character recognition. *Pattern Recognition (PR)*, 39(9):1575–1587, 2006. DOI: 10.1016/j.patcog.2006.03.011. 55, 58

Shibin Parameswaran and Kilian Q. Weinberger. Large Margin Multi-Task Metric Learning. In *Advances in Neural Information Processing Systems (NIPS) 23*, pages 1867–1875, 2010. 43

Kyoungup Park, Chunhua Shen, Zhihui Hao, and Junae Kim. Efficiently Learning a Distance Metric for Large Margin Nearest Neighbor Classification. In *Proceedings of the 25th AAAI Conference on Artificial Intelligence*, 2011. 22

Mateusz Pawlik and Nikolaus Augsten. RTED: a robust algorithm for the tree edit distance. *Proceedings of the VLDB Endowment*, 5(4):334–345, 2011. DOI: 10.14778/2095686.2095692. 13

Karl Pearson. On Lines and Planes of Closest Fit to Points in Space. *Philosophical Magazine*, 2 (6):559–572, 1901. DOI: 10.1080/14786440109462720. 14, 34

Michaël Perrot, Amaury Habrard, Damien Muselet, and Marc Sebban. Modeling perceptual color differences by local metric learning. In *Proceedings of the 13th European Conference on Computer Vision (ECCV)*, pages 96–111, 2014. DOI: 10.1007/978-3-319-10602-1_7. 92

Ali M. Qamar and Eric Gaussier. Online and Batch Learning of Generalized Cosine Similarities. In *Proceedings of the IEEE International Conference on Data Mining (ICDM)*, pages 926–931, 2009. DOI: 10.1109/ICDM.2009.114. 27

Ali M. Qamar, Eric Gaussier, Jean-Pierre Chevallet, and Joo-Hwee Lim. Similarity Learning for Nearest Neighbor Classification. In *Proceedings of the IEEE International Conference on Data Mining (ICDM)*, pages 983–988, 2008. DOI: 10.1109/ICDM.2008.81. 27, 74, 96

Guo-Jun Qi, Jinhui Tang, Zheng-Jun Zha, Tat-Seng Chua, and Hong-Jiang Zhang. An Efficient Sparse Metric Learning in High-Dimensional Space via l1-Penalized Log-Determinant Regularization. In *Proceedings of the 26th International Conference on Machine Learning (ICML)*, 2009. DOI: 10.1145/1553374.1553482. 24

Qi Qian, Rong Jin, Shenghuo Zhu, and Yuanqing Lin. An Integrated Framework for High Dimensional Distance Metric Learning and Its Application to Fine-Grained Visual Categorization. Technical report, arXiv:1402.0453, 2014. 32

Joaquin Quiñonero-Candela. *Dataset Shift in Machine Learning*. MIT Press, 2009. 48

Deva Ramanan and Simon Baker. Local Distance Functions: A Taxonomy, New Algorithms, and an Evaluation. *IEEE Transactions on Pattern Analysis and Machine Intelligence (TPAMI)*, 33(4):794–806, 2011. DOI: 10.1109/TPAMI.2010.127. 37

Chotirat Ann Ratanamahatana and Eamonn Keogh. Making time-series classification more accurate using learned constraints. In *Proceedings of the SIAM International Conference on Data Mining (SDM)*, pages 11–22, 2004. 59

Benjamin Recht, Christopher Re, Stephen J. Wright, and Feng Niu. Hogwild: A Lock-Free Approach to Parallelizing Stochastic Gradient Descent. In *Advances in Neural Information Processing Systems (NIPS) 24*, pages 693–701, 2011. 30

Eric S. Ristad and Peter N. Yianilos. Learning String-Edit Distance. *IEEE Transactions on Pattern Analysis and Machine Intelligence (TPAMI)*, 20(5):522–532, 1998. DOI: 10.1109/34.682181. 54, 55, 56, 58

Romer Rosales and Glenn Fung. Learning Sparse Metrics via Linear Programming. In *Proceedings of the 12th ACM SIGKDD International Conference on Knowledge Discovery and Data Mining*, pages 367–373, 2006. DOI: 10.1145/1150402.1150444. 25, 68, 73

Eleanor Rosch. Cognitive reference points. *Cognitive Psychology*, 7(4):532–547, 1975. DOI: 10.1016/0010-0285(75)90021-3. 26

Yossi Rubner, Carlo Tomasi, and Leonidas J. Guibas. The Earth Mover's Distance as a Metric for Image Retrieval. *International Journal of Computer Vision (IJCV)*, 40(2):99–121, 2000. DOI: 10.1023/A:1026543900054. 11, 51

Kate Saenko, Brian Kulis, Mario Fritz, and Trevor Darrell. Adapting Visual Category Models to New Domains. In *Proceedings of the 11th European Conference on Computer Vision (ECCV)*, pages 213–226, 2010. DOI: 10.1007/978-3-642-15561-1_16. 49

Hiroto Saigo, Jean-Philippe Vert, Nobuhisa Ueda, and Tatsuya Akutsu. Protein homology detection using string alignment kernels. *Bioinformatics*, 20(11):1682–1689, 2004. DOI: 10.1093/bioinformatics/bth141. 14, 56

Hiroto Saigo, Jean-Philippe Vert, and Tatsuya Akutsu. Optimizing amino acid substitution matrices with a local alignment kernel. *Bioinformatics*, 7(246):1–12, 2006. DOI: 10.1186/1471-2105-7-246. 56, 93

Hiroaki Sakoe. Dynamic programming algorithm optimization for spoken word recognition. *IEEE Transactions on Acoustics, Speech, and Signal Processing (TASS)*, 26:43–49, 1978. DOI: 10.1109/TASSP.1978.1163055. 59

Ruslan Salakhutdinov and Geoffrey E. Hinton. Learning a Nonlinear Embedding by Preserving Class Neighbourhood Structure. In *Proceedings of the 11th International Conference on Artificial Intelligence and Statistics (AISTATS)*, pages 412–419, 2007. 35

Gerard Salton, Andrew Wong, and C. S. Yang. A vector space model for automatic indexing. *Communications of the ACM*, 18(11):613–620, 1975. DOI: 10.1145/361219.361220. 10, 11, 15, 53

Craig Saunders, Alexander Gammerman, and Volodya Vovk. Ridge Regression Learning Algorithm in Dual Variables. In *Proceedings of the 15th International Conference on Machine Learning (ICML)*, pages 515–521, 1998. 14

Robert E. Schapire and Yoav Freund. *Boosting: Foundations and Algorithms*. MIT Press, 2012. 25

Jürgen Schmidhuber. Deep Learning in Neural Networks: An Overview. Technical report, arXiv:1404.7828, 2014. DOI: 10.1016/j.neunet.2014.09.003. 35

Bernhard Schölkopf and Alexander J. Smola. *Learning With Kernels, Support Vector Machines, Regularization, Optimization, and Beyond*. MIT Press, 2001. 1, 8, 14, 33

Bernhard Schölkopf, Alexander Smola, and Klaus-Robert Müller. Nonlinear component analysis as a kernel eigenvalue problem. *Neural Computation (NECO)*, 10(1):1299–1319, 1998. DOI: 10.1162/089976698300017467. 14, 34

Matthew Schultz and Thorsten Joachims. Learning a Distance Metric from Relative Comparisons. In *Advances in Neural Information Processing Systems (NIPS) 16*, 2003. DOI: 10.1145/1273496.1273523. 23, 32, 34, 74

Stanley M. Selkow. The tree-to-tree editing problem. *Information Processing Letters*, 6(6):184–186, 1977. DOI: 10.1016/0020-0190(77)90064-3. 13, 58

Shai Shalev-Shwartz and Tong Zhang. Stochastic Dual Coordinate Ascent Methods for Regularized Loss Minimization. *Journal of Machine Learning Research (JMLR)*, 14(1):567–599, 2013. 30

Shai Shalev-Shwartz, Yoram Singer, and Andrew Y. Ng. Online and batch learning of pseudo-metrics. In *Proceedings of the 21st International Conference on Machine Learning (ICML)*, 2004. DOI: 10.1145/1015330.1015376. 28, 34

Uri Shalit, Daphna Weinshall, and Gal Chechik. Online Learning in The Manifold of Low-Rank Matrices. In *Advances in Neural Information Processing Systems (NIPS) 23*, pages 2128–2136, 2010. 30

Uri Shalit, Daphna Weinshall, and Gal Chechik. Online Learning in the Embedded Manifold of Low-rank Matrices. *Journal of Machine Learning Research (JMLR)*, 13:429–458, 2012. 30

Blake Shaw, Bert C. Huang, and Tony Jebara. Learning a Distance Metric from a Network. In *Advances in Neural Information Processing Systems (NIPS) 24*, pages 1899–1907, 2011. 101

Chunhua Shen, Junae Kim, Lei Wang, and Anton van den Hengel. Positive Semidefinite Metric Learning with Boosting. In *Advances in Neural Information Processing Systems (NIPS) 22*, pages 1651–1660, 2009. 22, 25

Chunhua Shen, Junae Kim, Lei Wang, and Anton van den Hengel. Positive Semidefinite Metric Learning Using Boosting-like Algorithms. *Journal of Machine Learning Research (JMLR)*, 13: 1007–1036, 2012. 25

Roger N. Shepard. Toward a Universal Law of Generalization for Psychological Science. *Science*, 237(4820):1317–1323, 1987. DOI: 10.1126/science.3629243. 1

Yuan Shi, Aurélien Bellet, and Fei Sha. Sparse Compositional Metric Learning. In *Proceedings of the 28th AAAI Conference on Artificial Intelligence*, pages 2078–2084, 2014. 22, 32, 39, 68, 73

Robert D. Short and Keinosuke Fukunaga. The optimal distance measure for nearest neighbor classification. *IEEE Transactions on Information Theory (TIT)*, 27(5):622–626, 1981. DOI: 10.1109/TIT.1981.1056403. 1

Josef Sivic and Andrew Zisserman. Efficient visual search of videos cast as text retrieval. *IEEE Transactions on Pattern Analysis and Machine Intelligence (TPAMI)*, 31:591–606, 2009. DOI: 10.1109/TPAMI.2008.111. 10, 15, 27, 96

Temple F. Smith and Michael S. Waterman. Identification of common molecular subsequences. *Journal of Molecular Biology (JMB)*, 147(1):195–197, 1981. DOI: 10.1016/0022-2836(81)90087-5. 13

Suvrit Sra, Sebastian Nowozin, and Stephen J. Wright, editors. *Optimization for Machine Learning*. MIT Press, 2011. 3

Atsuhiro Takasu. Bayesian Similarity Model Estimation for Approximate Recognized Text Search. In *Proceedings of the 10th International Conference on Document Analysis and Recognition (ICDAR)*, pages 611–615, 2009. DOI: 10.1109/ICDAR.2009.193. 55

Daniel Tarlow, Kevin Swersky, Ilya Sutskever, Laurent Charlin, and Rich Zemel. Stochastic k-Neighborhood Selection for Supervised and Unsupervised Learning. In *Proceedings of the 30th International Conference on Machine Learning (ICML)*, 2013. 21

Lorenzo Torresani and Kuang-Chih Lee. Large Margin Component Analysis. In *Advances in Neural Information Processing Systems (NIPS) 19*, pages 1385–1392, 2006. 22, 34, 44

Ioannis Tsochantaridis, Thorsten Joachims, Thomas Hofmann, and Yasemin Altun. Large Margin Methods for Structured and Interdependent Output Variables. *Journal of Machine Learning Research*, 6:1453–1484, 2005. 46

Yuta Tsuboi, Hisashi Kashima, Shohei Hido, Steffen Bickel, and Masashi Sugiyama. Direct Density Ratio Estimation for Large-scale Covariate Shift Adaptation. In *Proceedings of the SIAM International Conference on Data Mining (SDM)*, pages 443–454, 2008. DOI: 10.2197/ipsjjip.17.138. 49

Amos Tversky. Features of similarity. *Psychological Review*, 84(4):327–352, 1977. DOI: 10.1037/0033-295X.84.4.327. 1, 26

Amos Tversky and Itamar Gati. Similarity, separability, and the triangle inequality. *Psychological Review*, 89(2):123–154, 1982. DOI: 10.1037/0033-295X.89.2.123. 26

Jeffrey K. Uhlmann. Satisfying general proximity / similarity queries with metric trees. *Information Processing Letters*, 40(4):175–179, 1991. DOI: 10.1016/0020-0190(91)90074-R. 26

Leslie G. Valiant. A theory of the learnable. *Communications of the ACM*, 27:1134–1142, 1984. DOI: 10.1145/1968.1972. 61, 64

Laurens van der Maaten and Geoffrey Hinton. Visualizing Data using t-SNE. *Journal of Machine Learning Research (JMLR)*, 9:2579–2605, 2008. 15

Laurens J.P. van der Maaten, Eric O. Postma, and H. Jaap van den Herik. Dimensionality Reduction: A Comparative Review. Technical report, Tilburg University, 2009. TiCC-TR 2009-005. 3

Aad W. van der Vaart and Jon A. Wellner. *Weak convergence and empirical processes.* Springer, 2000. 108

Vladimir N. Vapnik. *Statistical Learning Theory.* Wiley-Interscience, 1998. 82

Vladimir N. Vapnik and Alexey Y. Chervonenkis. On the uniform convergence of relative frequencies of events to their probabilities. *Theory of Probability and its Applications (TPA)*, 16(2): 264–280, 1971. DOI: 10.1137/1116025. 61

Raviteja Vemulapalli and David W. Jacobs. Riemannian Metric Learning for Symmetric Positive Definite Matrices. Technical report, arXiv:1501.02393, 2015. DOI: 10.1016/j.laa.2009.01.025. 101

Jarkko Venna, Jaakko Peltonen, Kristian Nybo, Helena Aidos, and Samuel Kaski. Information Retrieval Perspective to Nonlinear Dimensionality Reduction for Data Visualization. *Journal of Machine Learning Research (JMLR)*, 11:451–490, 2010. DOI: 10.1145/1756006.1756019. 15

Pascal Vincent, Hugo Larochelle, Yoshua Bengio, and Pierre-Antoine Manzagol. Extracting and composing robust features with denoising autoencoders. In *Proceedings of the 25th International Conference on Machine Learning (ICML)*, pages 1096–1103, 2008. DOI: 10.1145/1390156.1390294. 101

Kiri Wagstaff, Claire Cardie, Seth Rogers, and Stefan Schrödl. Constrained K-means Clustering with Background Knowledge. In *Proceedings of the 18th International Conference on Machine Learning (ICML)*, pages 577–584, 2001. 20

Fan Wang and Leonidas J. Guibas. Supervised Earth Mover's Distance Learning and Its Computer Vision Applications. In *Proceedings of the 12th European Conference on Computer Vision (ECCV)*, pages 442–455, 2012. DOI: 10.1007/978-3-642-33718-5_32. 51

Fang Wang, Shuqiang Jiang, Luis Herranz, and Qingming Huang. Improving image distance metric learning by embedding semantic relations. In *Proceedings of the 13th Pacific-Rim Conference on Multimedia (PCM)*, volume 7674, pages 424–434, 2012a. DOI: 10.1007/978-3-642-34778-8_39. 91

Jingyan Wang, Xin Gao, Quanquan Wang, and Yongping Li. ProDis-ContSHC: learning protein dissimilarity measures and hierarchical context coherently for protein-protein comparison in protein database retrieval. *BMC Bioinformatics*, 13(S-7):S2, 2012b. DOI: 10.1186/1471-2105-13-S7-S2. 95

Jun Wang, Huyen T. Do, Adam Woznica, and Alexandros Kalousis. Metric Learning with Multiple Kernels. In *Advances in Neural Information Processing Systems (NIPS) 24*, pages 1170–1178, 2011. 35

Jun Wang, Adam Woznica, and Alexandros Kalousis. Parametric Local Metric Learning for Nearest Neighbor Classification. In *Advances in Neural Information Processing Systems (NIPS) 25*, pages 1610–1618, 2012c. 22, 38

Qianying Wang, Pong C. Yuen, and Guocan Feng. Semi-supervised metric learning via topology preserving multiple semi-supervised assumptions. *Pattern Recognition (PR)*, 2013a. DOI: 10.1016/j.patcog.2013.02.015. 48

Yuyang Wang, Roni Khardon, Dmitry Pechyony, and Rosie Jones. Generalization Bounds for Online Learning Algorithms with Pairwise Loss Functions. In *Proceedings of the 25th Annual Conference on Learning Theory (COLT)*, pages 13.1–13.22, 2012d. 62

Yuyang Wang, Roni Khardon, Dmitry Pechyony, and Rosie Jones. Online Learning with Pairwise Loss Functions. Technical report, Tufts University, January 2013b. arXiv:1301.5332. 62

Zhengxiang Wang, Yiqun Hu, and Liang-Tien Chia. Image-to-class distance metric learning for image classification. In *Proceedings of the 10th European Conference on Computer Vision (ECCV)*, volume 6311, pages 706–719, 2010. DOI: 10.1007/978-3-642-15549-9_51. 89

Kilian Q. Weinberger and Lawrence K. Saul. Fast Solvers and Efficient Implementations for Distance Metric Learning. In *Proceedings of the 25th International Conference on Machine Learning (ICML)*, pages 1160–1167, 2008. DOI: 10.1145/1390156.1390302. 21, 38

Kilian Q. Weinberger and Lawrence K. Saul. Distance Metric Learning for Large Margin Nearest Neighbor Classification. *Journal of Machine Learning Research (JMLR)*, 10:207–244, 2009. DOI: 10.1145/1577069.1577078. 21, 38

Kilian Q. Weinberger, John Blitzer, and Lawrence K. Saul. Distance Metric Learning for Large Margin Nearest Neighbor Classification. In *Advances in Neural Information Processing Systems (NIPS) 18*, pages 1473–1480, 2005. DOI: 10.1145/1577069.1577078. 21

Lei Wu, Rong Jin, Steven C.-H. Hoi, Jianke Zhu, and Nenghai Yu. Learning Bregman Distance Functions and Its Application for Semi-Supervised Clustering. In *Advances in Neural Information Processing Systems (NIPS) 22*, pages 2089–2097, 2009. 40, 59

Lei Wu, Steven C.-H. Hoi, Rong Jin, Jianke Zhu, and Nenghai Yu. Learning Bregman Distance Functions for Semi-Supervised Clustering. *IEEE Transactions on Knowledge and Data Engineering (TKDE)*, 24(3):478–491, 2012. DOI: 10.1109/TKDE.2010.215. 40

Gaoyu Xiao and Anant Madabhushi. Aggregated distance metric learning (adm) for image classification in presence of limited training data. In *Proceedings of the 14th International Conference on Medical Image Computing and Computer-Assisted Intervention (MICCAI)*, volume 6893, pages 33–40, 2011. 91

Pengtao Xie and Eric Xing. Large Scale Distributed Distance Metric Learning. Technical report, arXiv:1412.5949, 2014. 31

Eric P. Xing, Andrew Y. Ng, Michael I. Jordan, and Stuart J. Russell. Distance Metric Learning with Application to Clustering with Side-Information. In *Advances in Neural Information Processing Systems (NIPS) 15*, pages 505–512, 2002. 1, 19

Caiming Xiong, David Johnson, Ran Xu, and Jason J. Corso. Random forests for metric learning with implicit pairwise position dependence. In *Proceedings of the 18th ACM SIGKDD International Conference on Knowledge Discovery and Data Mining*, pages 958–966, 2012. DOI: 10.1145/2339530.2339680. 41

Huilin Xiong and Xue-Wen Chen. Kernel-based distance metric learning for microarray data classification. *BMC Bioinformatics*, 7:299, 2006. DOI: 10.1186/1471-2105-7-299. 95

Huan Xu and Shie Mannor. Robustness and Generalization. In *Proceedings of the 23rd Annual Conference on Learning Theory (COLT)*, pages 503–515, 2010. DOI: 10.1007/s10994-011-5268-1. 68

Huan Xu and Shie Mannor. Robustness and Generalization. *Machine Learning Journal (MLJ)*, 86(3):391–423, 2012. DOI: 10.1007/s10994-011-5268-1. 62, 68, 69

Huan Xu, Constantine Caramanis, and Shie Mannor. Sparse Algorithms Are Not Stable: A No-Free-Lunch Theorem. *IEEE Transactions on Pattern Analysis and Machine Intelligence (TPAMI)*, 34(1):187–193, 2012. DOI: 10.1109/TPAMI.2011.177. 68

Rui Xu and Don C. Wunsch. *Clustering*. Wiley - IEEE Press, 2008. 15

Liu Yang and Rong Jin. An efficient algorithm for local distance metric learning. In *Proceedings of the 20th AAAI Conference on Artificial Intelligence*, 2006. 91

Liu Yang, Rong Jin, Lily B. Mummert, Rahul Sukthankar, Adam Goode, Bin Zheng, Steven C. H. Hoi, and Mahadev Satyanarayanan. A boosting framework for visuality-preserving distance metric learning and its application to medical image retrieval. *IEEE Transactions on Pattern Analysis and Machine Intelligence (TPAMI)*, 32(1):30–44, 2010. DOI: 10.1109/TPAMI.2008.273. 91

Peipei Yang, Kaizhu Huang, and Cheng-Lin Liu. Multi-Task Low-Rank Metric Learning Based on Common Subspace. In *Proceedings of the 18th International Conference on Neural Information Processing (ICONIP)*, pages 151–159, 2011. DOI: 10.1007/978-3-642-24958-7_18. 44

Peipei Yang, Kaizhu Huang, and Cheng-Lin Liu. Geometry Preserving Multi-task Metric Learning. In *Proceedings of the European Conference on Machine Learning and Principles and Practice of Knowledge Discovery in Databases (ECML/PKDD)*, pages 648–664, 2012. DOI: 10.1007/s10994-013-5379-y. 44

Rui Yang, Panos Kalnis, and Anthony K. H. Tung. Similarity evaluation on tree-structured data. In *Proceedings of the ACM SIGMOD International Conference on Management of Data (COMAD)*, pages 754–765, 2005. DOI: 10.1145/1066157.1066243. 13

Wen-Tau Yih, Kristina Toutanova, John Platt, and Chris Meek. Learning discriminative projections for text similarity measures. In *Proceedings of the 15th Conference on Computational Natural Language Learning (CoNNL)*, 2011. 97

Yiming Ying and Peng Li. Distance Metric Learning with Eigenvalue Optimization. *Journal of Machine Learning Research (JMLR)*, 13:1–26, 2012. 20

Yiming Ying, Kaizhu Huang, and Colin Campbell. Sparse Metric Learning via Smooth Optimization. In *Advances in Neural Information Processing Systems (NIPS) 22*, pages 2214–2222, 2009. 25, 68, 73, 74

Daren Yu, Xiao Yu, Qinghua Hu, Jinfu Liu, and Anqi Wu. Dynamic time warping constraint learning for large margin nearest neighbor classification. *Information Sciences*, 181(13):2787–2796, 2011. DOI: 10.1016/j.ins.2011.03.001. 59

Zheng-Jun Zha, Tao Mei, Meng Wang, Zengfu Wang, and Xian-Sheng Hua. Robust Distance Metric Learning with Auxiliary Knowledge. In *Proceedings of the 21st International Joint Conference on Artificial Intelligence (IJCAI)*, pages 1327–1332, 2009. 48

De-Chuan Zhan, Ming Li, Yu-Feng Li, and Zhi-Hua Zhou. Learning instance specific distances using metric propagation. In *Proceedings of the 26th International Conference on Machine Learning*, 2009. DOI: 10.1145/1553374.1553530. 38

Changshui Zhang, Feiping Nie, and Shiming Xiang. A general kernelization framework for learning algorithms based on kernel PCA. *Neurocomputing*, 73(4–6):959–967, 2010. DOI: 10.1016/j.neucom.2009.08.014. 34

Kaizhong Zhang and Dennis Shasha. Simple fast algorithms for the editing distance between trees and related problems. *SIAM Journal of Computing (SICOMP)*, 18(6):1245–1262, 1989. DOI: 10.1137/0218082. 13, 58

Yu Zhang and Dit-Yan Yeung. Transfer metric learning by learning task relationships. In *Proceedings of the 16th ACM SIGKDD International Conference on Knowledge Discovery and Data Mining*, pages 1199–1208, 2010. DOI: 10.1145/1835804.1835954. 45

Guoqiang Zhong, Kaizhu Huang, and Cheng-Lin Liu. Low Rank Metric Learning with Manifold Regularization. In *Proceedings of the IEEE International Conference on Data Mining (ICDM)*, pages 1266–1271, 2011. DOI: 10.1109/ICDM.2011.95. 48

Ji Zhu, Saharon Rosset, Trevor Hastie, and Robert Tibshirani. 1-norm Support Vector Machines. In *Advances in Neural Information Processing Systems (NIPS) 16*, pages 49–56, 2003. 77

Authors' Biographies

AURÉLIEN BELLET

Aurélien Bellet received his Ph.D. in Machine Learning from the University of Saint-Etienne (France) in 2012. His work focused on algorithmic and theoretical aspects of metric and similarity learning. After completing his thesis, he was a postdoctoral researcher at the University of Southern California, where he worked on large-scale and distributed machine learning with applications to automatic speech recognition. He is currently a postdoctoral researcher at Télécom ParisTech (France), working on machine learning for big data.

AMAURY HABRARD

Amaury Habrard received a Ph.D. in Machine Learning in 2004 from the University of Saint-Etienne. He was Assistant Professor at the Laboratoire d'Informatique Fondamentale of Aix-Marseille University until 2011, where he received a habilitation thesis in 2010. He is currently Professor in the Machine Learning group at the Hubert Curien laboratory of the University of Saint-Etienne. His research interests include metric learning, transfer learning, online learning and learning theory.

MARC SEBBAN

Marc Sebban received a Ph.D. in Machine Learning in 1996 from the Université of Lyon 1. After four years spent at the French West Indies and Guyana University as Assistant Professor, he got a position of Professor in 2002 at the University of Saint-Etienne (France). Since 2010, he is the head of the Machine Learning group and the director of the Computer Science, Cryptography and Imaging department of the Hubert Curien laboratory. His research interests focus on ensemble methods, metric learning, transfer learning and more generally on statistical learning theory.

Printed in the United States
by Baker & Taylor Publisher Services